ENGINEERING APPLICATIONS OF BIOCHAR

Edited by **Wu-Jang Huang**

Engineering Applications of Biochar
http://dx.doi.org/10.5772/65231
Edited by Wu-Jang Huang

Contributors

Ahmed Abdelhafez, Mohamed Abass, Li Jianhua, Ki Young Park, Daegi Kim, Daike Tian, Amei Liu, Yanci Xiang, Tao Zhang, Wu-Jang Huang

Published by InTech

Janeza Trdine 9, 51000 Rijeka, Croatia

Publishing Process Manager Marijana Francetic
Technical Editor SPi Global
Cover InTech Design team

First published August, 2017
Printed in Croatia
Legal deposit, Croatia: National and University Library in Zagreb

Additional hard copies can be obtained from orders@intechopen.com

Engineering Applications of Biochar, Edited by Wu-Jang Huang
p. cm.
Print ISBN 978-953-51-3403-9
Online ISBN 978-953-51-3404-6

For EU product safety concerns:
IN TECH d.o.o., Prolaz Marije Krucifikse Kozulić 3, 51000 Rijeka, Croatia,
info@intechopen.com or visit our website at intechopen.com.

Meet the editor

Wu-Jang Huang, PhD, is a full professor of the Department of Environmental Engineering and Science of National Pingtung University of Science and Technology. He has coauthored over 80 publications and advised over 60 master and 10 doctoral program students. He was approved to this position at 35 years of age and is one of the 100 youngest full professors in Taiwan at 2010. His research interests are actually focused on the material science and environmental science and engineering. Synthesis, characterization, recycling, and novel materialization of solid wastes and their toxicity investigation for people's future life are all in considerations.

Contents

Preface

The discovery of *biochar* has opened a new research field to connect the oldest carbon material with traditional farming technique. Expected for the agriculture applications, the engineered biochar materials have attracted more attention in the recent years, especially for the *zero waste* production of agriculture, throughout the world. Through self-heating, different benefits can be gained by converting such wastes into functional biochar materials than biological conversion.

This book has included four sections, including the "Introduction", "History of Biochar," "Preparation of Biochar," and "Applications of Biochar." The editor would like to thank all the authors who contributed their valuable works. We hope that the development of biochar can cross its application field from agriculture into engineering.

Prof. Wu-Jang Huang
Department of Environmental Engineering and Science (EES)
National Pingtung University of Science and Technology (NPUST)
Pingtung, Taiwan

Section 1

Introduction

Chapter 1

Introductory Chapter: How We Could Use Biochar in Engineering

Wu-Jang Huang

Additional information is available at the end of the chapter

http://dx.doi.org/10.5772/intechopen.69480

1. Introduction

The application of biochar has been shown to improve soil physical properties, especially in soils with bad soil structure or high bulk density. Biochar is made from a pyrolysis process that occurs spontaneously at the temperatures that **are usually below 700°C**. At its most extreme state, pyrolysis leaves only a carbon residue, which is called carbonization. The high temperatures used in pyrolysis induce polymerization of the molecules within the feedstocks, producing larger molecules and thermal decomposition of some feedstock components into smaller molecules. The remaining solid component following pyrolysis is charcoal, referred to as biochar, when produced with the intention of using it for soil improvement [2, 4, 5]. Basically, *biochar is known as a pyrolyzed carbon from solid waste used in agriculture since 1998.*

The use of burned residues in farming field has been carried for long lone time. The most near is Williams et al. [6], who discussed the advantages and drawbacks of burning versus incorporating straw in rice growing. Due to the rice husk, biochar has high silica (SiO_2) contents, and silicon (Si) is a beneficial element for plant growth that helps plants overcome multiple stresses including biotic and abiotic stresses. The benefits of silicon in crop production are, therefore, healthier plants and higher yield with fewer applications of pesticides and other chemical products [1]. Our recent studies also indicated that the application of rice husk, bamboo and woody biochar can change the physical properties on soil, and we also found that an application rate of 10 t ha^{-1} should not be exceeded when applying biochar on these soils [3]. Expected for the agriculture applications, the engineered biochar materials have attracted more attentions in recent years, especially for the ZERO WASTES production of agriculture in the world. Through self-heating, to convert such wastes into functional biochar materials can gain different benefits than that of biological conversion. **The editor hopes that the development of biochar can cross its application field from agriculture into engineering.**

Author details

Wu-Jang Huang

Address all correspondence to: wjhaung@mail.npust.edu.tw

Department of Environmental Engineering and Science (EES), National Pingtung University of Science and Technology (NPUTC), Pingtung, Taiwan

References

[1] Heckman JR. The Soil Profile. Rutgers Cooperative Extension, Plant Biology and Pathology Department, University of New Jersey. Vol. 20. 2012. Available from: http://njaes.rutgers.edu/pubs/soilprofile/sp-v20.pdf

[2] Husain L, Khan A, Shareef A, Ahmed T. Forest Fire Derived Black Carbon in the Adirondack Mountains. NY; 2008. pp. 1745-1850

[3] Milla OV, Rivera EB, Huang W-J, Chien C-C, Wang Y-M. Agronomic properties and characterization of rice husk and wood biochars and their effect on the growth of water spinach in a field test. Journal of Plant Nutrition and Soil Science. 2013;**13**(2):251-266

[4] Preston C, Schmidt M. Black (pyrogenic) carbon in boreal forests: A synthesis of current knowledge and uncertainties. Biogeosciences Discussions. 2006;**3**:211-271

[5] Schmidt M, Skjemstad J, Gehrt E, Kögel-Knabner I. Charred organic carbon in German chernozemic soils. The European Journal of Soil Science. 1999;**50**:351-365

[6] Williams NA, Morse ND, Buckman JF. Burning vs. incorporation of rice crop residues. Agronomy Journal. 1972;**64**:467-468

Section 2

History of Biochar

Chapter 2

Biochar: The Black Diamond for Soil Sustainability, Contamination Control and Agricultural Production

Ahmed A. Abdelhafez, Mohamed H.H. Abbas and Jianhua Li

Additional information is available at the end of the chapter

http://dx.doi.org/10.5772/intechopen.68803

Abstract

Production of biochars from agricultural wastes reduces significantly the volume and weight of the wastes, and hence, it can be considered as a promising means for managing the agricultural wastes. Biochar has received great interest during the last few years, due to its beneficial role to mitigate CO_2 emission through enhancing the long-term carbon sequestration. The effects of biochar on soil properties vary widely, depending on the characteristics of soil and the biochar. Most types of biochars are of alkaline nature and of high C content. Addition of biochar to the soil can improve the cation exchange capacity enrich soil with the nutrients and enhance the microbial growth, and improve some soil physical properties such as water retention and aggregation. For contamination control, biochars have proven to be a suitable tool for controlling the contaminants in the environment. The high surface area, porous structure, alkaline nature, and the presence of functional groups characterized the biochar as alternative option for the remediation of heavy metal contaminated waters and soils. However, there is a lack of knowledge regarding the effects of biochar in the presence of mineral and/or organic fertilizers on the plant growth and nutrient transformation in soils. In addition, biochar is successfully used for treating the acid soils; therefore, future studies are needed to investigate the neutralization of alkaline performance of biochar to be used safely in alkaline soils.

Keywords: biochar, soil, mineral fertilizers, contaminants, environmental management

1. Introduction

The increasing demand for food and the fertilizers (inorganic and organic) is a day fact. Mineral fertilizers are of great importance for food production. Accordingly, the world demand for mineral fertilizers has increased during the last few decades to meet the increased demand for agricultural products in response to the growing population [1]. However, several studies indicated that mineral fertilizers contain various amounts of heavy metals as impurities [1, 2]. Consequently, the addition of large amounts of mineral fertilizers to the soil for long periods may result in the accumulation of heavy metals in soils and plant grown thereon, as confirmed by previous studies [1, 3].

Intensive researches have been performed to overcome the infertility problems of the agricultural soils. The addition of organic residues in forms of compost, manure, and other organic forms has proven to be suitable alternative options for mineral fertilization. However, these materials need to be applied intensively due to their low nutritive contents and rapid degradation rate. Biochar is a charcoal produced from the pyrolysis of biomass at relatively low temperature (<700°C) [4]. Biochar has received great interest during the last few years, due to its beneficial roles in environmental management. Several beneficial objectives could be achieved through biochar application for environment, that is, waste management, energy production, soil improvement, maximizing agricultural production, contamination control (soils and waters), and greenhouse gases (GHGs) mitigation [5–10].

2. Production and characterization of biochars derived from different wastes

2.1. Historical view of the biochar

Biochar production is an ancient practice over that past 70 centuries in the Egyptian societies. It seems that the production of biochar was not the main target, the Egyptian societies used the liquid wood tars to embalm the bodies of their dead, and the liquid preserving agent was produced from charring processes [11]. Similarly, the use of biochar as soil amendment first began over the past 2,500 years in South America (terra preta), the place which named "the black earth." Biochar is created both naturally by forest fires and by human through burning bits for different practices, that is, cooking and manufacturing. Terra preta is a famous soil located in the Amazon Basin. The acidic condition of terra preta in the past due to the toxic levels of exchangeable aluminum hindered the agricultural production; however, the continuous accumulation of biochar in the soils led to enrich the soil in calcium and phosphate and elevated pH level in comparison with the surrounding soils. In addition, terra preta soil contains about 50 Mg ha^{-1} carbon in a form of biochar within approximately 1 m depth [12]. Consequently, aluminum toxicity in this soil was neutralized, and soil status in terms of physical, biological, and chemical features has been modified that made it one of the most fertile soils over the world. The promising benefits of biochar have alerted the sign for researchers in the past to determine the positive performance of biochar, for example, the role of biochar for improving vegetative growth and enhancing soil fertility has been studied by Trimble [13] and Retan [14]. Due to the several benefits of biochar, many researches and extension initiatives of biochar have been established all over the world in order to spread the knowledge and cooperation of biochar and its applications, for example, the Australia New Zealand Biochar Research Network (www.anzbiochar.org/project.html), the US Biochar Initiative (http://biochar-us.org/biochar-research), the European Biochar Research Network (http://cost.european-biochar.org/en), the UK Biochar Research Center (http://www.biochar.ac.uk/), the China Biochar Network (http://www.biochar-international.org/chinanetwork), the Japan Biochar Association (http://www.geocities.jp/yasizato/JBA.htm), the New Zealand Biochar Research Centre (http://www.massey.ac.nz/massey/learning/colleges/college-of-sciences/research/agriculture-environment-research/soil-earth-sciences/biochar-research-centre/biochar-research-centre_home.cfm) and the Biochar India (www.biocharindia.com).

2.2. Production of biochar

2.2.1. *Biomasses for biochar production*

The rapid population growth led to subsequent increases in food production, and consequently, large amounts of organic residues are produced annually [8]. Therefore, it is essential to recycle their organic residues effectively. Various types of biomass have been used for biochar production, including: (i) agricultural and forestry by-products, that is, wood chips, straw, nut shells, rice hulls, tree bark, wood pellets, and switch grass, (ii) industrial by-products, that is, sugar cane bagasse, paper sludge, and pulp, (iii) animal wastes such as chicken litter, dairy and swine manure, and (iv) sewage sludge. Producing the biochar from biomass, especially wastes offer an excellent way for the recycling of wastes into beneficial materials. Pyrolysis treatment reduces the volume of biomasses by 44–90 and 75–80% and weight by 44–93 and 71–77% [8, 15].

2.2.2. *Production technologies of biochar*

Biochar is produced through the pyrolysis process, in which the biomasses are burned in the absence of oxygen. As mentioned above, the main objective of biochar production is to use it as a soil amendment or for usage in other aspects such as remediation and industrial technologies. The process is closely similar to those of gasification; however, in case of gasification, the process is performed in two steps, firstly, the biomass is heated to around 600°C, and hydrocarbon gases and tar are evaporated; secondly, char is gasified by reaction with oxygen, hydrogen, and steam under high temperature. However, in case of pyrolysis, the biomass is burned in the absence of oxygen along the production time. There are many important secondary products upon producing the biochar, including a synthetic gas that can be used to generate electricity and bio-oil, which can be used as diesel fuel. As shown in **Table 1**, biochar can be produced through fast and slow pyrolysis techniques; the main difference between them is the heating rate and the amount of the produced bio-oil.

2.2.3. *Development of biochar production*

Figure 1 shows the development of biochar production. The people used to simply gather piles of agricultural wastes and cover them and burn them slowly with limited air. They have

Parameter	Biochar production		
	Fast pyrolysis	Slow pyrolysis	Gasification
Temperature	~500°C	~400°C	600–1800°C
Heating rate	up to 1000°C min^{-1}	Slow 5–30°C min^{-1}	–
Time	Few seconds	Hours ~ days	–
Aeration	Oxygen free	Oxygen free or limited	Oxygen limited
Biochar	~12%	~35%	~10%
Syngas	~13%	~35%	~85%
Bio-oil	~75%	~30%	~5%

Data obtained from Roos [16].

Table 1. Differences between gasification and pyrolysis processes.

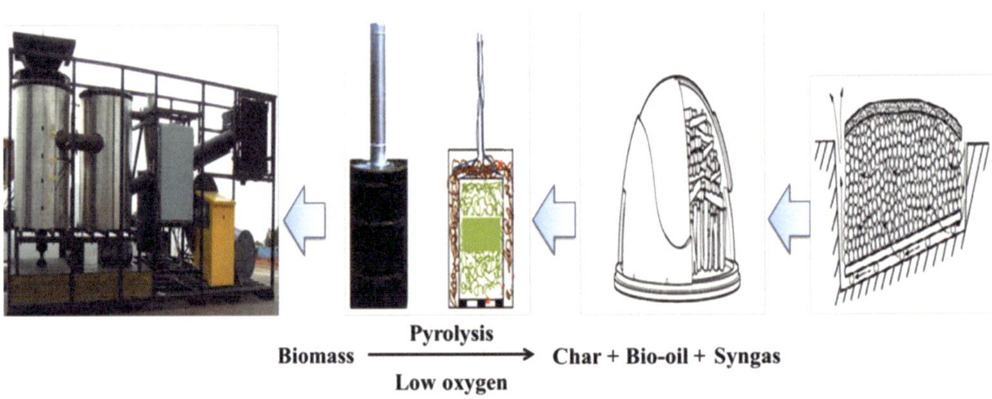

Figure 1. Development of biochar production. Images obtained from Lehmann and Joseph [4], Abdelhafez et al. [7].

used several ways to exclude air penetration into burning places, such like covering with soil particles. This traditional method is still used today in developing countries; however, considerable amounts of smokes and almost half amount of carbon dioxide in the original biomass are released into the atmosphere. Briefly, biomasses were put together tightly and covered with a layer of soil in a large pit kiln then a small part of the biomass was burned up. To achieve a successful pyrolysis process, people used to make small holes in the soil surface to provide amount of air uniformly in order to maintain a productive balance between burning and pyrolysis. The pit kiln has some disadvantages, that is, the release of almost 50% of C into the atmosphere and the high ash content of the produced biochar. To overcome these problems, brick kilns were developed to achieve more control for aeration. These kilns were better insulated and allowed a better airflow control, which allowed higher biochar yields and lower ash contents of the produced biochar. The above-mentioned techniques are in situ biochar production units, where the biochar was made at places where suitable raw material was abundant. By beginning of the 1930s, transportable, cylindrical metal kilns were developed in Europe and became popular in the 1960s, in developing countries. They are often made out of oil drums and are more easily to handle than traditional pits. The sealed container allows a high control of airflow, and the biochar can easier be recovered [17]. The portable kilns are still used in developing countries in the small farms and have been used experimentally by Abdelhafez et al. [8, 9] in China and Egypt, respectively. However, the traditional methods may contaminate the environment due to the emitted syngas and bio-oils. Therefore, advanced instruments have been developed successfully to eliminate the emitted syngas and bio-oil and to use them as by products by using specific condensers for gas and bio-oil collection.

2.3. Differences between biochar and charcoal

Man used to create charcoal instinctively for heating, industry and production beginning from the creation. Both biochar and charcoal contain high carbon materials; however, there are some major differences as follows [4]:

(i) Charcoal is produced primarily as a source of energy, while biochar is manufactured as a soil amendment for improving soil fertility, carbon (C) sink, or water filtration.

(ii) Wood is the major source of charcoal production; however, biochar can be produced from any biomass.

(iii) The carbonization trend of biochar is not complete as in charcoal; consequently, charcoal contains much ash content compared to the biochar.

Carbon is present in the biochar in a form of six C atoms linked together. The formation of graphite is more likely to occur when the C atoms arranged together without O or H ions. However, in case of biochar, graphite does not form because the arranged atoms of carbon are corrupted by O and H ions; as a result, C atoms are arranged irregularly according to the type of biomass used for biochar production and temperature of pyrolysis [4].

2.4. Physicochemical characteristics of biochar

All biochars are black but are not created equal and are not of the same physicochemical characteristics. Both the types of biomass and pyrolysis conditions play important roles for identifying the characteristics of the produced biochars [5, 18]. The produced material of biochar is a solid, structured, carbonaceous material and exhibits a high surface area [19], low oxygen and hydrogen contents [20], and little amount of nutrients [21, 22]. The physical characteristics of the produced biochar depend mainly on the type of biomass and the pyrolysis conditions, in terms of, heating rate, highest temperature of burning, pressure, burning time and the characteristics of burning vessel. It is well known that organic materials start to decompose after 120°C; hemicellulose compounds decompose at 200–260°C, and lignins decompose at 240–350°C [23]. Biochar has proven to be a suitable tool for the removal of heavy metals from aqueous solutions [10] due to the presence of macrospores with an average pore size of 51–138 m^2 g^{-1} [24],. The presence of functional groups on the surface of biochar candidate it for the removal of organic and inorganic contaminants from aqueous solutions. Abdelhafez and Li [10] demonstrated that the spectrums of sugar cane and orange peel biochars are quite similar; both biochars exhibited absorption bands on 3448.13 and 3429.4 cm^{-1} corresponding to C—OH functional groups; around 1637.27 and 1384.85 cm^{-1}, there were C=O and C—C bands and the adsorption bands on 1101.43 cm^{-1} present the C—O, C—C, and C—OH bands. Therefore, both biochars had the ability to adsorb Pb(II) ions from aqueous solutions. During the pyrolysis of biomass, heating causes some nutrients to be volatilized, especially at the surface of the material, while other nutrients become concentrated in the remaining biochar. In case of wood-rich materials, carbon (C) begins to volatilize around 100°C, N above 200°C, S above 375°C, and potassium (K) and P between 700°C and 800°C. The volatilization of magnesium (Mg), calcium (Ca), and manganese (Mn) occurs at temperatures above 1000°C [25, 26]. Therefore, biochar contains much amount of alkali metal ions causing its liming performance when it is applied to the soils [8, 9]. As shown in **Table 2**, more than 80% of the produced biochars is C, while nitrogen contents are relatively low because most of nitrogen in the feedstock starts to be volatile at temperature above 200°C. Therefore, the nitrogen contents of biochars derived from agricultural wastes are quite low. However, the nitrogen content of sewage sludge biochar seems to be higher than the agricultural wastes biochars [15]. Furthermore, most of the stated biochars characterized by its high pH values, and this could be attributed to the presence of alkaline metal ions, that is, Ca, Mg, and K, which are stable and does not volatile in the biomass during the production of biochars. The previous studies demonstrated that increasing the pyrolysis time and temperature led to increase the surface area and pours structure of the produced biochar [27, 28]. Similarly, the pH of the produced biochar depends on the pyrolysis temperature and time; by increasing the pyrolysis temperature, the pH of the produced biochars increased to reach 11.5 in some studies [29]. A point to note that, biochar has a liming effect when it is applied to the soil; therefore, possible increment in soil acidity (pH) might occur [8]. In addition, adsorption of macronutrients (N, P and K) on the surfaces of biochar might hinder its uptake by the growing plants. Applying biochar to the soils has been found to increase the bioavailability and plant uptake of phosphorus (P), alkaline metals and some trace metals [30], but the mechanisms for these increases are still a matter of speculation. Moreover, the benefits of biochar for the removal of organic and inorganic contaminants from water are well documented [31, 32]. However, to date, only limited studies are available on biochar effects combined with different mineral and organic fertilization levels on soil properties and plant growth. The behavior of biochar is not equal for all elements; some studies have reported that biochar has the potential for the stabilization of Pb in shooting range and metal smelter contaminated soils [7, 10]. Abdelhafez et al. [7, 8]. illustrated the beneficial effect of biochar for soil improvement and Pb remediation in a military shooting range and metal smelter contaminated soils. Moreover, it was found that biochar increased the bioavailability of Cu (shooting range soil) and As (metal smelter soil). Therefore, the chemical behavior of biochar with heavy metal ions is not constant and needs to be investigated.

Feedstock	Temperature	pH	%							CEC, cmolc kg^{-1}	C/N ratio	%		H/C ratio	O/C ratio	SSA, $m^2 g^{-1}$	Reference
			C	N	P	S	Ca	Mg	K			O.M	Ash				
Peanut hull	500.00	8.60	82.00	2.70	0.30	0.10	–	–	–	–	30.37	–	9.30	0.44	0.03	200.00	[5]
Sugar cane bagasse	<500	8.63	74.02	1.00	0.24	–	0.17	0.32	2.00	69.62	74.02	87.80	12.21	0.42	0.23	92.30	[10]
Orange peel	<500	8.75	66.36	2.13	0.25	–	1.04	0.28	1.86	68.28	31.15	88.80	11.17	0.65	0.32	0.20	
Cattle waste	380.00	8.20	62.10	0.10	–	–	–	–	–	39.00	621.00	–	25.60	1.90	0.27	–	[15]
Sewage sludge	380.00	8.50	38.30	5.20	–	–	–	–	–	0.50	7.37	–	44.90	0.94	0.25	–	
Oak wood	600.00	6.38	87.50	0.20	–	–	–	–	–	75.70	489.00	–	0.01	0.33	0.07	642.00	[33]
Corn stover	350.00	9.39	60.40	1.20	–	–	–	–	–	419.30	51.00	–	11.40	0.75	0.29	293.00	[34]
	600.00	9.42	70.60	1.07	–	–	–	–	–	252.10	66.00	–	16.70	0.39	0.10	527.00	
Corn stalk	400.00	9.60	51.10	1.34	0.25	–	–	–	1.34	–	38.13	–	–	–	–	–	[35]
	500.00	10.10	48.40	0.55	0.44	–	–	–	2.65	–	88.00	–	–	–	–	–	
Wheat straw	425.00	10.40	46.70	0.59	–	–	1.00	0.60	2.60	–	79.15	–	20.80	–	–	–	[36]
Coco peat	500.00	10.30	84.40	1.02	0.03	0.27	0.06	2.30	–	–	82.75	–	15.90	0.41	0.10	13.70	[37]
Coconut charcoal	<500	8.86	76.50	0.20	–	–	–	–	–	–	426.60	–	2.90	0.12	–	–	[38]
Pine wood	<500	8.47	53.20	0.40	–	–	–	–	–	–	143.40	–	65.70	0.35	–	–	
Eucalyptus deglupta	350.00	7.00	82.40	0.57	0.06	0.03	–	–	–	4.69	144.56	–	0.20	–	0.12		[39]
Hard wood saw dust	500.00	–	63.80	0.22	–	0.01	–	–	–		290.00	–	22.80	0.60	0.14	1.00	[40]
Chinese pine	600.00	8.38	66.67	2.21	–	–	–	–	–	31.58	30.17	–	12.50	0.58	0.31	–	[41]

Table 2. Physicochemical characteristics of different types of biochar.

3. The beneficial effect of biochars derived from different wastes on soil fertility

3.1. Fresh organic matter versus biochar as soil amendment

Soil organic carbon is originated by photosynthesis under highly reduced conditions (estimated by 600 mV) which are presented in leaf chloroplast [42]. Such fresh materials are probably the most reduced fraction when added to soils, acting as electron pumps to more oxidized species [43]. Generally, organic residues are used as amendments to improve soil quality and productivity [44].

The organic amendments that persist longer in soil might exert high impacts on soil physicochemical reactivity [45]. In deep soil layers, organic materials are relatively more stable than in the surface ones due to the absence of fresh organic carbon, an essential source of energy for soil microbes [46]. Probably, compounds that contain less oxygen (lower electron richness) are less easily decomposed than do compounds having comparable size, solubility, and molecular complexity [47].

Specific mechanisms might guarantee stabilization of organic C in soil, for example, biotic exclusion which might take place through adsorption of organics and aggregation with soil minerals forming mineral-bound OM [48]. Also, preservation of recalcitrant (stable) compounds might stabilize organic C [49]. It is thought that the recalcitrant compounds are present in organic materials in much higher proportions than those classified as labile [50].

Pyrolysis is the converting of unstable organic matter into more stable forms (biochar) that can be applied to soils [51]. This can be attained by heating carbon bearing solids in the absence of oxygen [52] to produce porous materials of low density [53] and more stable forms of carbon [54] which are more resistant to biodegradation as compared to fresh organic materials. Biochars' half-life in soil is relatively long [55].

3.2. Effect of biochar on soil properties

Biochar is used as an amendment to improve soil properties. It improves soil-water holding capacity [56, 57], saturated hydraulic conductivity [58], increases cation exchange capacity (CEC) [8, 59], decreases bulk density [60], and minimizes the loss of nutrients and other agricultural chemicals in soil run-off [4]. It also decreases soil penetration resistance and increases aggregation and infiltration [61]. On the other hand, biochar does not show any significant effect on soil porosity either directly through pore contribution, or indirectly through improving aggregate stability [62]. Besides, applications of biochar increase soil electrical conductivity (EC) in addition to its high contents of phosphorus and nitrogen [63].

3.3. Effect of biochar on the plant growth and soil biota

The main roles of biochar for enhancing plant growth are directly through its nutrients contents, and indirectly through its effects on nutrients use efficiency. Biochar serves as energy provider [64] for wheat [65], rice [36], maize [6], soy bean [66], and spring barley [67]; thus, it improves root density, crop growth, and productivity [68]. It was found that chicken manure-derived biochar increased the dry weights of the shoot and root of the Indian mustard by 353 and 572% upon its application to soil at a rate of only 1% [69]. Even biochar produced from wastewater sludge increased the productivity of cherry tomatoes by 64% as compared to the control [63]. Thus, such amendment is recommended for low-fertile and degraded soils [4] as well as highly weathered soil [70]. The zone of plant rhizosphere becomes larger with application of biochar [71]. Moreover, biochar increases plant resistance toward biotic stresses [72]. Some types of biochar amendments are rich in nutrients [73], and on the other hand, it minimizes the leaching of nutrients from soil, i.e., nitrate [74], ammonium, and probably phosphate [75]. However, the majority of biochars produce ethylene which is an inhibitor for soil microbes [68], beside of the released organic molecules which might suppress activities of some beneficial soil biota [76].

4. Biochar: alternative option for soil sustainability

Using biochar as a soil amendment can fulfill three main targets, that is, increasing plant productivity, thus achieving food security [4], improving soil properties, and disputing land degradation [77] beside of minimizing the change of climate [78]. Moreover, biochar changes organic wastes into value-added biochar which acts as sorbents for eliminating contaminants in wastewater [79]. As mentioned above, the transformation of terra preta soil into a high fertile soil due to biochar addition is a great evidence of the role of biochar for soil sustainability. The recycling of agricultural wastes into benefit materials guarantees the sustainability of agricultural lands.

5. Biochar and the environmental change

5.1. Effect of biochar on CO_2 emissions

Soils can store more carbon than do plants or atmosphere [47]. Globally, soil organic matter (SOM) contains about three times as much carbon as either the atmosphere or terrestrial vegetation [80]. In soils of low N content, CO_2 is the dominant greenhouse gases (GHGs) component [81]. Accordingly, strategies that migrate excess CO_2 from atmospheric air might be more important than reducing equivalent emissions of CO_2 to air [64]. The promising approach in lowering CO_2 from air is biochar [78]. Thus, biochar could be considered as the geo-engineering solution to control climate change [82] probably by means of carbon sequestration [83], thus minimizing the emissions of the greenhouse gases [84] while supplying energy and improving the productivity of the cultivated crops [64]. Roberts et al. [84] found that 62–66% of CO_2 emissions could be sequestered within biochar. Accordingly, adopting biochar technologies can offer financial incentive in emission trading markets [82].

5.2. Effect of biochar on CH_4 and N_2O emissions

Pyrolysis process serves also in reducing emissions of the other GHGs such as methane (CH_4) and nitrous oxide (N_2O) when amended to agricultural soils and pastures [64]. Biochar decreases the emissions of CH_4 and, therefore, increases the stock of soil organic carbon [85]. This probably takes place through suppressing the oxidation of ambient CH_4 [51]. On the other hand, the emissions of CH_4 might increase in rice paddy soil amended with biochar [36].

The effect of biochar on the transformation processes of nitrogen (N) in soil is not well defined [86]. Probably, biochar reduced GHGs emissions only in neutral to acidic soils with high N content [87]. In this concern, emissions of N_2O as well as leaching ammonium from soil could be reduced when using biochar rather than fresh organic material as soil amendments [86]. Generally, biochar suppresses production of N_2O [54]. It is found that 10.7–41.8% of the total emissions of N_2O decreased with application of biochar at rates of 20 and 40 Mg ha^{-1}, respectively [87]. Similar results show that soil N_2O fluxes decreased up to 79% in soils amended with biochar as compared to the control [88]. In an experiment conducted by Mukherjee et al. [89], it was found that 92% of the cumulative N_2O emissions reduced when amending soils with biochar. Even under the reduced conditions of the rice paddy soil, biochar can also minimize the emission of N_2O [36]. Such reductions might be attributed to the oxidative reactions that take place on the surfaces of biochar with ageing [86]. Accordingly, reductions of the emissions of N_2O owing to application of biochar to soils improve the GHGs-to-yield ratio conditions [90].

Others found no significant differences in emissions of both CO_2 and N_2O from soils owing to application of biochar as compared to nonamended soils [56]. Likewise, Mukherjee et al. [89] found that the total cumulative emissions of CH_4 and CO_2 emissions were not affected significantly by amending soils with biochar. It is worthy to mention that biochar production itself can increase, to some extent, the greenhouse gases emitted to the atmosphere; however, more studies are needed to fulfill this point of study and to lessen GHGs emitted during production process.

6. The beneficial role of biochar for contamination control of soils and waters

6.1. Biochar as means for decontaminating soils from heavy metals and pesticides

Biochars produced at relatively high temperature pyrolysis are more efficient in sorption of organic contaminants, whereas those produced at low temperatures are more efficient for removing heavy metals [102]. At low temperature, the produced biochar is of acidic nature, whereas those produced at high temperature were of alkaline nature [91]. This approach offers a new safe solution for decontaminating soil pollution [92]. Generally, biochars are efficient in reducing the phytoavailability of many organic pollutants in soil, that is, (1) herbicides, for example, atrazine and acetochlor [51], Fluometuron and 4-chloro-2-methylphenoxyacetic acid [93], (2) pesticides, for example, pyrimethanil [94], atrazine [95], simazine [96], azoxystrobin [97], (3) fungicides, that is, tricyclazole in alluvial paddy soil [98] in addition to (4) phenols [99], thus controlling their toxicity and transfer in soil [100]. Immobilization of these organic residues might be take place because of the high affinity and ability of biochar to sequester such organics [101]. High temperature pyrolysis biochar is characterized by its high surface area, high micro-porosity, and hydrophobicity [102], and thus, combined adsorption and partition mechanisms might take place with the herbicide, pesticides, and the fungicide on carbonized and noncarbonized fractions [96]. In case of phenols, its sorption might take place on the microspores surface area of the biochar in addition to sorption on the carboxylic and lactonic groups [99]. Sorption affinity with the organic contaminants is found irreversible [94] and can increase with decreasing solid/solution ratio [96].

Biochars can also immobilize the phytotoxcity of heavy metals in soil forming less bioavailable organic bound fraction [69]. Biochar is of an alkaline nature, thus applying biochar to soils is associated with increases in soil pH [103]. The mechanism of immobilization might be a result of precipitation due to the rise in soil pH due to the application of the basic biochar or even by the electrostatic interaction on the carboxyl groups of the biochar [104] or through coordination by π electrons (C=C) of carbon [105].

Many experiments revealed the successfulness of biochar treatments on partitioning of heavy metals in soil, for example, Cd, Cu, and Pb [69]. Surprisingly, using biochar for decontaminating soils decreased the leachable fractions of Cd and Zn by 300 and 45-folds in compared to the untreated treatments [106]. In another experiment, it was found that treating soils with biochar removed Pb, Zn, and Cd by 97.4, 53.4, and 54.5%, respectively [107]. It is worthy to mention that the oxidized biochars, rich in carboxyl groups, showed higher affinity to immobilize Pb, Cu, and Zn than did the un-oxidized ones [104].

6.2. Biochar as a means of decontaminating heavy metals and organic residues from wastewater

Biochars act as sorbents for decontaminating wastewaters from heavy metals [79]. This might take place mainly through sorption on the surface functional groups of biochar [108], for example, oxygen-containing carboxyl, hydroxyl, and phenolic surface functional groups [109]. The kinetics of adsorption followed pseudo second order [10, 110]. The stability of heavy metals by biochar correlated significantly with the oxygen-containing functional groups of the biochar [108] with maximum adsorption attained within the pH range 5.0–6.0 [110]. Digested dairy waste biochar and digested whole sugar beet biochar were found to be efficient in removing Pb^{2+}, Cu^{2+}, Ni^{2+}, and Cd^{2+} from wastewater [111]. Also, biochars can efficiently remove organic contaminants from wastewaters. It was found that the fast pyrolysis pine wood biochar could remove salicylic acid and ibuprofen from solutions [112].

Biochar can also effectively remove phosphate from wastewater [113]. This probably takes place on the colloidal and nano-sized MgO particles on its surface [114]. Most of the sorbed phosphate is bioavailable and can be added to soils as slow release P-fertilizers [115]. Moreover, 60% of the sorbed phosphate can be desorbed within 24 h [116].

Treating biochar hydrothermally with H_2O_2 increased its affinity to remove heavy metals from aqueous solutions because this treatment increased the oxygen-containing functional groups [117]. Another type of biochar is chitosan-modified one which is a low-cost synthesized biochar efficient for immobilizing heavy metal in the environment [118]. Also, a graphene/biochar composite is a safe economic adsorbent that can decontaminate heavy metals through surface complexation with C—O, C C, —OH, and O C— groups [119].

7. Conclusion and future challenges of biochar

The previous demonstration showed that biochar plays an important role in environmental management and soil sustainability. Several beneficial roles of biochar have been observed. Biochar improves soil fertility and plant growth, mitigates the greenhouse gasses, and could be used successfully for the remediation of soils and waters from contaminants. However, several research questions are still unknown and need intensive researches, that is, the effect of biochar on minerals and/or organic fertilizers use efficiency and the neutralization of alkaline performance of biochar to be used safely in alkaline soils. In addition, the stability of biochar in the amended soils needs a sustainable experiment to determine exactly the degradation rate of different types of biochars.

Author details

Ahmed A. Abdelhafez[1]*, Mohamed H.H. Abbas[2] and Jianhua Li[3]

*Address all correspondence to: ahmed.aziz@aun.edu.eg; leejianhua@tongji.edu.cn

1 Department of Soils and Water, Faculty of Agriculture, Assiut University (New Valley Branch), Egypt

2 Department of Soils and Water, Faculty of Agriculture, Benha University, Egypt

3 Key Laboratory of Yangtze River Water Environment, Ministry of Education, College of Environmental Science and Engineering, Tongji University, Shanghai, China

References

[1] Abdelhafez AA, Abbas HH, Abd-El-Aal RS, Kandi NF, Li J, Mahmoud W. Environmental and health impacts of successive mineral fertilization in Egypt. Clean-Soil, Air and Water. 2012;**40**(4):356-363

[2] Milinovic J, Lukic V, Nikolic-Mandic S, Stojanovic D. Concentrations of heavy metals in NPK fertilizers imported in Serbia. Pesticide Phytomedicine. 2008;**23**:195-200

[3] Huang S, Jin J, He P. Effects of different patterns of land use on status of heavy metals in agricultural soils. Better Crops. 2009;**93**:20-22

[4] Lehmann J, Joseph S. Biochar for environmental management: An introduction. In: Lehmann J, Joseph S. (Eds.), Biochar for Environmental Management: Science and Technology. London: Earthscan; 2009. pp. 1-12

[5] Novak JM, Lima I, Xing B, Gaskin JW, Steiner C, Das KC, Ahmedna M, Rehrah D, Watts DW, Busscher WJ, Schomberg H. Characterization of designer biochar produced at

different temperatures and their effects on a loamy sand. Annals of Environmental Science. 2009;**3**(1):195-206

[6] Major J, Rondon M, Molina D, Riha SJ, Lehmann J. Maize yield and nutrition during 4 years after biochar application to a Colombian savanna oxisol. Plant and Soil. 2010;**333**:117-128

[7] Abdelhafez AA, Lee SS, Ok YS. Effects of biochar on soil quality and heavy metal bioavailability in a military shooting range soil. Korean Society of Soil and Ground Water Environment (KOSSGE). 2010; pp. 236. Korea

[8] Abdelhafez AA, Li J, Abbas MHH. Feasibility of biochar manufactured from organic wastes on the stabilization of heavy metals in a metal smelter contaminated soil. Chemosphere. 2014;**117**:66-71

[9] Abdelhafez AA, Abbas MHH, Hamed MH. Biochar: A solution for soil Pb pollution. The 8th International Conference for Development and the Environment in the Arab World Assiut University, Egypt. 2016; pp. 89-103

[10] Abdelhafez AA, Li J. Removal of Pb(II) from aqueous solution by using biochars derived from sugarcane bagasse and orange peel. Journal of the Taiwan Institute of Chemical Engineers. 2016;**61**:367-375

[11] Emrich W. Handbook of charcoal making. The traditional and industrial methods. D. Reidel Publishing Company - Dordrecht, Holland; 1985

[12] Verheijen FGA, Jeffery S, Bastos AC, van der Velde M, Diafas I. Biochar Application to Soils—A Critical Scientific Review of Effects on Soil Properties, Processes and Functions. EUR 24099 EN. Luxembourg: Office for the Official Publications of the European Communities; 2009. 149 p

[13] Trimble WH. On charring wood. Plough, the Loom and the Anvil. 1851;**3**:513-516

[14] Retan GA. Charcoal as a means of solving some nursery problems. Forestry Quarterly. 1915;**13**:25-30

[15] Shinogi Y, Yoshida H, Koizumi T, Yamaoka M, Saito T. Basic characteristics of low-temperature carbon products from waste sludge. Advances in Environmental Research. 2003;**7**(3):661-665

[16] Roos CJ. Clean Heat and Power Using Biomass Gasification for Industrial and Agricultural Projects, U.S. Department of Energy Olympia; 2010. ss: 1-9

[17] FAO Forestry Department, editor. Simple Technologies for Charcoal Making. 2nd ed. Rome: FAO; 1987

[18] Chen B. Yuan M. Enhanced sorption of polycyclic aromatic hydrocarbons by soil amended with biochar. Journal of Soils and Sediments. 2011;**11**:62-71

[19] Bird MI, Ascough PL, Young IM, Wood CV, Scott AC. X-ray microtomographic imaging of charcoal. Journal of Archaeological Science. 2008;**35**:2698-2706

[20] Abdullah H, Wu H. Biochar as a fuel: 1. Properties and grindability of biochars produced from the pyrolysis of mallee wood under slow-heating conditions. Energy and Fuels. 2009;**23**:4174-4181

[21] Gaskin JW, Steiner C, Harris K, Das KC, Bibens B. Effect of low temperature pyrolysis conditions on biochar for agricultural use. Transactions of the ASABE. 2008;**51**:2061-2069

[22] Agblevor FA, Beis S, Kim SS, Tarrant R, Mante NO. Biocrude oils from the fast pyrolysis of poultry litter and hardwood. Waste Management. 2010;**30**:298-307

[23] Sjöström E. Wood Chemistry: Fundamentals and Applications. 2nd ed. San Diego, CA: Academic Press; 1993

[24] Laine J, Yunes S. Effect of the preparation method on the pore size distribution of activated carbon from coconut shell. Carbon. 1992;**30**:601-604

[25] Neary DG, Klopatek CC, DeBano LF, Folliott PF. Fire effects on belowground sustainability: A review and synthesis. Forest Ecology and Management. 1999;**122**:51-71

[26] Knoepp JD, DeBano LF, Neary DG. Soil Chemistry, RMRS-GTR 42-4. Ogden, UT: US Department of Agriculture, Forest Service, Rocky Mountain Research Station; 2005

[27] Keiluweit M, Kleber M, Sparrow MA, Simoneit BR, Prahl FG. Solvent-extractable polycyclic aromatic hydrocarbons in biochar: Influence of pyrolysis temperature and feedstock. Environmental Science and Technology. 2012;**46**(17):9333-9341

[28] Bird MI, Wurster CM, de Paula Silva PH, Bass AM, de Nys R. Algal biochar—Production and properties. Bioresource Technology. 2011;**102**(2):1886-1891

[29] Yuan JH, Xu RK, Zhang H. The forms of alkalis in the biochar produced from crop residues at different temperatures. Bioresource Technology. 2011;**102**(3):3488-3497

[30] Houben D, Evrard L, Sonnet P. Mobility, bioavailability and pH-dependent leaching of cadmium, zinc and lead in a contaminated soil amended with biochar, Chemosphere. 2013;**92**:1450-1457

[31] Liu XH, Zhang XC. Effect of biochar on pH of alkaline soils in the loess plateau: Results from incubation experiments. International Journal of Agriculture and Biology. 2012;**14**:745-750

[32] Pellera FM, Giannis A, Kalderis D, Anastasiadou K, Stegmann R, Wang J-Y, Gidarakos E. Adsorption of Cu(II) ions from aqueous solutions on biochars prepared from agricultural by-products. Journal of Environmental Management. 2012;**96**:35-42

[33] Nguyen B, Lehmann J. Black carbon decomposition under varying water regimes. Organic Geochemistry. 2009;**40**:846-853

[34] Nguyen B, Lehmann J, Hockaday WC, Joseph S, Masiello CA. Temperature sensitivity of black carbon decomposition and oxidation. Environmental Science and Technology. 2010;**44**:3324-3331

[35] Feng Y, Xu Y, Yu Y, Xie Z, Lin X. Mechanisms of biochar decreasing methane emission from Chinese paddy soils. Soil Biology and Biochemistry. 2012;**46**:80-88

[36] Zhang A, Cui L, Pan G, Li L, Hussain Q, Zhang X, Zheng J, Crowley D. Effect of biochar amendment on yield and methane and nitrous oxide emissions from a rice paddy from Tai Lake plain, China. Agriculture, Ecosystems and Environment. 2010;**139**:469-475

[37] Lee Y, Park J, Ryu C, Gang KS, Yang W, Park Y, Jung J, Hyun S. Comparison of biochar properties from biomass residues produced by slow pyrolysis at 500°C. Bioresource Technology. 2013;**148**:196-201

[38] Erin N, Yargicoglu EN, Sadasivam BY, Reddy KR, Spokas K. Physical and chemical characterization of waste wood derived biochars. Waste Management. 2014;**36**:256-268

[39] Rondon MA, Lehmann J, Ramírez J, Hurtado M. Biological nitrogen fixation by common beans (*Phaseolus vulgaris* L.) increases with biochar additions. Biology and Fertility of Soils. 2007;**43**(6):699-708

[40] Fabbri D, Torri C. Spokas KA. Analytical pyrolysis of synthetic chars derived from biomass with potential agronomic application (biochar): Relationships with impacts on microbial carbon dioxide production. Journal of Analytical and Applied Pyrolysis. 2012;**93**:77-84

[41] Liu XH, Han FP, Zhang XC. Effect of biochar on soil aggregates in the loess plateau: Results from incubation experiments. International Journal of Agriculture & Biology. 2012;**14**:975-979

[42] Macías F, Camps Arbestain M. Soil carbon sequestration in a changing global environment. Mitigation and Adaptation Strategies for Global Change. 2010;**15**:511-529

[43] Chesworth W. Redox, soils and carbon sequestration. Edafologia. 2004;**11**:37-43

[44] Melero S, Porras JCR, Herencia JF, Madejon E. Chemical and biochemical properties in a silty loam soil under conventional and organic management. Soil and Tillage Research. 2006;**90**:162-170

[45] Wander M. Soil Organic Matter Fractions and their Relevance to Soil Function. Soil Organic Matter in Sustainable Agriculture. CRC Press USA, Fred Magdoff and Ray R. Weil; 2004

[46] Fontaine S, Barot S, Barre P, Bdioui N, Mary B, Rumpel C. Stability of organic carbon in deep soil layers controlled by fresh carbon supply. Nature. 2007;**450**:277-280

[47] Kleber M. What is recalcitrant soil organic matter. Environmental Chemistry. 2010;**7**: 320-332

[48] Lützow M, Kögel-Knabner I, Ekschmitt K, Matzner E, Guggenberger G, Marschner B, Flessa H. Stabilization of organic matter in temperate soils: Mechanisms and their relevance under different soil conditions—A review. European Journal of Soil Science. 2006;**57**:426-445

[49] Lützow M, Kögel-Knabner I, Ludwig B, Matzner E, Flessa H, Ekschmitt K, Guggenberger G, Marschner B, Kalbitz K. Stabilization mechanisms of organic matter in four temperate soils: Development and application of a conceptual model. Journal of Plant Nutrition and Soil Science. 2008;**171**:111-124

[50] Davidson EA, Janssens IA. Temperature sensitivity of soil carbon decomposition and feedbacks to climate change. Nature. 2006;**440**:165-173

[51] Spokas KA, Koskinena WC, Bakera JM, Reicoskyb DC. Impacts of woodchip biochar additions on greenhouse gas production and sorption/degradation of two herbicides in a Minnesota soil. Chemosphere. 2009;**77**:574-581

[52] Brown R. Biochar production technology. In: Lehmann J, Joseph S, editors. Biochar for Environmental Management: Science and Technology. London: Earthscan; 2009. pp. 127-146

[53] Beesley L, Moreno-Jiménez E, Gomez-Eyles JL, Harris E, Robinson B, Sizmur T. A review of biochars' potential role in the remediation, revegetation and restoration of contaminated soils. Environmental Pollution. 2011;**159**:3269-3282

[54] Spokas KA, Reicoky DC. Impacts of sixteen different biochars on soil greenhouse gas production. Annals of Environmental Science. 2009;**3**:179-193

[55] Verheijen FGA, Montanarella L, Bastos AC. Sustainability, certification, and regulation of biochar. Pesquisa Agropecuária Brasileira. 2012;**47**:649-653

[56] Karhu K, Mattila T, Bergström I, Regina K. Biochar addition to agricultural soil increased CH_4 uptake and water holding capacity—Results from a short-term pilot field study. Agriculture, Ecosystems & Environment. 2011;**140**:309-313

[57] Lal AMAR. Biochar impacts on soil physical properties and greenhouse gas emissions. Agronomy. 2013;**3**:313-339

[58] Asai H, Samson BK, Stephan HM, Songyikhangsuthor K, Homma K, Kiyono Y, Inoue Y, Shiraiwa T, Horie T. Biochar amendment techniques for upland rice production in Northern Laos: 1. Soil physical properties, leaf SPAD and grain yield. Field Crops Research. 2009;**111**:81-84

[59] Martin SL, Clarke ML, Othman M, Ramsden SJ, West HM. Biochar-mediated reductions in greenhouse gas emissions from soil amended with anaerobic digestates. Biomass and Bioenergy. 2015;**79**:39-49

[60] Mukherjee A, Lal R. Biochar impacts on soil physical properties and greenhouse gas emissions. Agronomy. 2013;**3**:313-339

[61] Busscher WJ, Novak JM, Evans DE, Watts DW, Niandou MAS, Ahmedna M. Influence of pecan biochar on physical properties of a Norfolk loamy sand. Soil Science. 2010;**175**:10-14

[62] Hardie M, Clothier B, Bound S, Oliver G. Close D. Does biochar influence soil physical properties and soil water availability?. Plant and Soil. 2014;**376**:347-361

[63] Hossain MK, Strezov V, Yin Chan K, Nelson PF. Agronomic properties of wastewater sludge biochar and bioavailability of metals in production of cherry tomato (*Lycopersicon esculentum*). Chemosphere. 2010;**78**:1167-1171

[64] Woolf D, Amonette JE, Street-Perrott FA, Lehmann J, Joseph S. Sustainable biochar to mitigate global climate change. Nature Communications 2010;**1**(56):1-9

[65] Baronti S, Alberti G, Delle Vedove G, Di Gennaro F, Fellet G, Genesio L, Miglietta F, Peressotti A, Vaccari FP. The Biochar option to improve plant yields: First results from some field and pot experiments in Italy. Italian Journal of Agronomy. 2010;**5**:3-11

[66] Mete FZ, Mia S, Dijkstra FA, Abuyusuf M, Hossain A.S.M.I. Synergistic effects of biochar and NPK fertilizer on soybean yield in an alkaline soil. Pedosphere. 2015;**25**:713-719

[67] Sun Z, Bruun EW, Arthur E, de Jonge LW, Moldrup P, Hauggaard-Nielsen H, Elsgaard L. Effect of biochar on aerobic processes, enzyme activity, and crop yields in two sandy loam soils. Biology and Fertility of Soils. 2014;**50**:1087-1097

[68] Spokas KA, Baker JM, Reicosky DC. Ethylene: Potential key for biochar amendment impacts. Plant and Soil. 2010;**333**:443-452

[69] Park JH, Choppala GK, Bolan NS, Chung JW, Chuasavathi T. Biochar reduces the bioavailability and phytotoxicity of heavy metals. Plant and Soil. 2011;348-439

[70] Lehmann J, Rondon M. Bio-char soil nanagement on highly weathered soils in the humid tropics. Biological Approaches to Sustainable Soil Systems. CRC Press -Taylor and Francis Group, LLC; 2006. pp. 517-529

[71] Prendergast-Miller MT, Duvall M, Sohi SP. Biochar-root interactions are mediated by biochar nutrient content and impacts on soil nutrient availability. European Journal of Soil Science. 2014;**65**:173-185

[72] Elad Y, Cytryn E, Harel YM, Lew B, Graber ER. The biochar effect: Plant resistance to biotic stresses. Phytopathologia Mediterranea. 2012;**50**:335-349

[73] Uchimiya M, Lima IM, Klasson KT, Wartelle LH. Contaminant immobilization and nutrient release by biochar soil amendment: Roles of natural organic matter. Chemosphere. 2010b;**80**:935-940

[74] Dempster DN, Jones DL, Murphy DV. Clay and biochar amendments decreased inorganic but not dissolved organic nitrogen leaching in soil. Soil Research. 2012;**50**:216-221

[75] Yao Y, Gao B, Zhang M, Inyang M, Zimmerman AR. Effect of biochar amendment on sorption and leaching of nitrate, ammonium, and phosphate in a sandy soil. Chemosphere. 2012;**89**:1467-1471

[76] Lehmann J, Rillig MC, Thies J, Masiello CA, Hockaday WC, Crowley D. Biochar effects on soil biota—A review. Soil Biology and Biochemistry. 2011;**43**:1812-1836

[77] Barrow CJ. Biochar: Potential for countering land degradation and for improving agriculture. Applied Geography. 2012;**34**:21-28

[78] Lehmann J. Bio-energy in the black. Frontiers in Ecology and the Environment. 2007;**5**:381-387

[79] Xu X, Cao X, Zhao L. Comparison of rice husk-and dairy manure-derived biochars for simultaneously removing heavy metals from aqueous solutions: Role of mineral components in biochars. Chemosphere. 2013;**92**:955-961

[80] Schmidt MWI, Torn MS, Abiven S, Dittmar T, Guggenberger G, Janssens IA, Kleber M, Kogel-Knabner I, Lehmann J, Manning DAC, Nannipieri PRasse DP, Weiner S, Trumbore SE. Persistence of soil organic matter as an ecosystem property. Nature. 2011;**478**:49-56

[81] Zheng J, Stewart CE, Cotrufo MF. Biochar and nitrogen fertilizer alters soil nitrogen dynamics and greenhouse gas fluxes from two temperate soils. Journal of Environmental Quality. 2012;**41**:1361-1370

[82] Cowie AL, Downie AE, George BH, Singh B-P, Van Zwieten L, O'Connell D. Is sustainability certification for biochar the answer to environmental risks? Pesquisa Agropecuária Brasileira. 2012;**47**:637-648

[83] Lehmann J, Gaunt J, Rondon M. Bio-char sequestration in terrestrial ecosystems—A review. Mitigation and Adaptation Strategies for Global Change. 2006;**11**:395-419

[84] Roberts KG, Gloy BA, Joseph S, Scott NR, Lehmann J. Life cycle assessment of biochar systems: Estimating the energetic, economic, and climate change potential. Environmental Science and Technology. 2010;**44**:827-833

[85] Xie Z, Xu Y, Liu G, Liu Q, Zhu J, Tu C, Amonette JE, Cadisch G, Yong JWH, Hu S. Impact of biochar application on nitrogen nutrition of rice, greenhouse-gas emissions and soil organic carbon dynamics in two paddy soils of China. Plant and Soil. 2013;**370**:527-540

[86] Singh BP, Hatton BJ, Singh B, Cowie AL, Kathuria A. Influence of biochars on nitrous oxide emission and nitrogen leaching from two contrasting soils. Journal of Environmental Quality. 2010;**39**:1224-1235

[87] Zhang A, Liu Y, Pan G, Hussain Q, Li L, Zheng J, Zhang X. Effect of biochar amendment on maize yield and greenhouse gas emissions from a soil organic carbon poor calcareous loamy soil from Central China plain. Plant and Soil. 2012;**351**:263-275

[88] Castaldi S, Riondino M, Baronti S, Esposito FR, Marzaioli R, Rutigliano FA, Vaccari FP, Miglietta F. Impact of biochar application to a Mediterranean wheat crop on soil microbial activity and greenhouse gas fluxes. Chemosphere. 2011;**85**:1464-1471

[89] Mukherjee A, Lal R, Zimmerman AR. Effects of biochar and other amendments on the physical properties and greenhouse gas emissions of an artificially degraded soil. Science of the Total Environment. 2014;**487**:26-36

[90] Kammann C, Ratering S, Eckhard C, Müller C. Biochar and hydrochar effects on greenhouse gas (carbon dioxide, nitrous oxide, and methane) fluxes from soils. Journal of Environmental Quality. 2012;**41**:1052-1066

[91] Hossain MK, Strezov V, Chan KY, Ziolkowski A, Nelson PF. Influence of pyrolysis temperature on production and nutrient properties of wastewater sludge biochar. Journal of Environmental Management. 2011;**92**:223-228

[92] Zhang X, Wang H, He L, Lu K, Sarmah A, Li J, Bolan NS, Pei J, Huang H. Using biochar for remediation of soils contaminated with heavy metals and organic pollutants. Environmental Science and Pollution Research. 2013;**20**:8472-8483

[93] Cabrera A, Cox L, Spokas KA, Celis R, Hermosín MC, Cornejo J, Koskinen WC. Comparative sorption and leaching study of the herbicides fluometuron and 4-chloro-2-methylphenoxyacetic acid (MCPA) in a soil amended with biochars and other sorbents. Journal of Agricultural and Food Chemistry. 2011;**59**:12550-12560

[94] Yu X, Pan L, Ying G, Kookana RS. Enhanced and irreversible sorption of pesticide pyrimethanil by soil amended with biochars. Journal of Environmental Sciences. 2010;**22**:615-620

[95] Cao X, Ma L, Gao B, Harris W. Dairy-manure derived biochar effectively sorbs lead and atrazine. Environmental Science and Technology. 2009;**43**:3285-3291

[96] Zheng W, Guo M, Chow T, Bennett DN, Rajagopalan N. Sorption properties of greenwaste biochar for two triazine pesticides. Journal of Hazardous Materials. 2010;**181**:121-126

[97] Sopeña F, Bending GD. Impacts of biochar on bioavailability of the fungicide azoxystrobin: A comparison of the effect on biodegradation rate and toxicity to the fungal community. Chemosphere. 2013;**91**:1525-1533

[98] García-Jaramillo M, Cox L, Knicker HE, Cornejo J, Spokas KA, Hermosín, MC. Characterization and selection of biochar for an efficient retention of tricyclazole in a flooded alluvial paddy soil. Journal of Hazardous Materials. 2015;**286**:581-588

[99] Han Y, Boateng AA, Qi PX, Lima IM, Chang J. Heavy metal and phenol adsorptive properties of biochars from pyrolyzed switchgrass and woody biomass in correlation with surface properties. Journal of Environmental Management. 2013;**118**:196-204.

[100] Smernik RJ. Biochar and sorption of organic compounds. In: Lehmann J, Joseph S, editors. Biochar for Environmental Management: Science and Technology. London: Earthscan; 2009. pp. 289-300

[101] Yu X, Ying G, Kookana RS. Reduced plant uptake of pesticides with biochar additions to soil. Chemosphere. 2009;**76**:665-671

[102] Ahmad M, Rajapaksha AU, Lim JE, Zhang M, Bolan N, Mohan D, Vithanage M, Lee SS, Ok YS. Biochar as a sorbent for contaminant management in soil and water: A review. Chemosphere. 2014;**99**:19-33

[103] Peng X, Ye LL, Wang C, Zhou H, Sun B. Temperature and duration-dependent rice straw-derived biochar: Characteristics and its effects on soil properties of an ultisol in southern China. Soil Tillage Research. 2011;**112**(2):159-166

[104] Uchimiya M, Bannon DI, Wartelle LH. Retention of Heavy Metals by Carboxyl Functional Groups of Biochars in Small Arms Range Soil. Journal of Agricultural and Food Chemistry. 2012;**60**:1798-1809

[105] Uchimiya L, Lima IM, Klasson KT, Chang S, Wartelle LH, Rodgers JE. Immobilization of heavy metal ions (CuII, CdII, NiII, and PbII) by Broiler litter-derived biochars in water and soil. Journal of Agricultural and Food Chemistry. 2010;**58**:5538-5544

[106] Beesley L, Marmiroli M. The immobilisation and retention of soluble arsenic, cadmium and zinc by biochar. Environmental Pollution. 2011;**159**:474-480

[107] Liang Y, Cao X, Zhao L, Arellano E. Biochar- and phosphate-induced immobilization of heavy metals in contaminated soil and water: Implication on simultaneous remediation of contaminated soil and groundwater. Environmental Science and Pollution Research. 2014;**21**:4665-4674

[108] Uchimiya M, Wartelle LH, Klasson KT, Fortier CA, Lima IM, Influence of pyrolysis temperature on biochar property and function as a heavy metal sorbent in soil. Journal of Agricultural and Food Chemistry. 2011;**59**:2501-2510

[109] Uchimiya M, Chang S, Klasson KT. Screening biochars for heavy metal retention in soil: Role of oxygen functional groups. Journal of Hazardous Materials. 2011;190:432-441

[110] Kołodyńska D, Wnętrzak R, Leahy JJ, Hayes MHB, Kwapiński W, Hubicki Z. Kinetic and adsorptive characterization of biochar in metal ions removal. Chemical Engineering Journal. 2012;**197**:295-305

[111] Inyang M, Gao B, Yao Y, Xue Y, Zimmerman AR, Pullammanappallil P, Cao X. Removal of heavy metals from aqueous solution by biochars derived from anaerobically digested biomass. Bioresource Technology. 2012;**110**:50-56

[112] Essandoh M, Kunwar B, Pittman Jr CU, Mohan D, Mlsna T. Sorptive removal of salicylic acid and ibuprofen from aqueous solutions using pine wood fast pyrolysis biochar. Chemical Engineering Journal. 2015;**265**:219-227

[113] Chen B, Chen Z, Lv S. A novel magnetic biochar efficiently sorbs organic pollutants and phosphate. Bioresource Technology. 2011;**102**:716-723

[114] Yao Y, Gao B, Inyang M, Zimmerman AR, Cao X, Pullammanappallil P, Yang L. Removal of phosphate from aqueous solution by biochar derived from anaerobically digested sugar beet tailings. Journal of Hazardous Materials. 2011;**190**:501-507

[115] Yao Y, Gao B, Chen J, Yang L. Engineered biochar reclaiming phosphate from aqueous solutions: Mechanisms and potential application as a slow-release fertilizer. Environmental Science and Technology. 2013;**47**:8700-8708

[116] Sarkhot DV, Ghezzehei TA, Berhe AA. Effectiveness of biochar for sorption of ammonium and phosphate from dairy effluent. Journal of Environmental Quality. 2013;**42**:1545-1554

[117] Xue Y, Gao B, Yao Y, Inyang M, Zhang M, Zimmerman AR, Ro KS. Hydrogen peroxide modification enhances the ability of biochar (hydrochar) produced from hydrothermal carbonization of peanut hull to remove aqueous heavy metals: Batch and column tests. Chemical Engineering Journal. 2012;**200**:673-680

[118] Zhou Y, Gao B, Zimmerman AR, Fang J, Sun Y, Cao X. Sorption of heavy metals on chitosan-modified biochars and its biological effects. Chemical Engineering Journal. 2013;**231**:512-518

[119] Tang J, Lv H, Gong Y, Huang Y. Preparation and characterization of a novel graphene/biochar composite for aqueous phenanthrene and mercury removal. Bioresource Technology. 2015;**196**:355-363

Section 3

Preparation of Biochar

Chapter 3

Conversion of Municipal Solid Wastes into Biochar through Hydrothermal Carbonization

Daegi Kim, Ki Young Park and Kunio Yoshikawa

Additional information is available at the end of the chapter

http://dx.doi.org/10.5772/intechopen.68221

Abstract

In this study, the hydrothermal treatment of municipal solid wastes (MSWs) for the production of biochar as a renewable solid fuel was investigated. The properties of surrogate MSWs and mixtures of newspaper and vegetables were greatly improved by hydrothermal treatment and were similar to those of coal-like fuel substances. Hydrothermal treatment increased the calorific value, the fixed carbon, and carbon contents. The composition of the major biomass components of MSW was found to affect the alternation of their physical and chemical properties significantly. These characteristic changes in pure cellulose, hemicellulose, and lignin were similar to those of coalification at the hydrothermal reaction temperature range of 150–280°C. The treated products became a solid fuel substance, the characteristics of which corresponded with fuel between lignite and subbituminous coal. The results of this study indicate that hydrothermal treatment can be used as an effective means to generate highly energy-efficient renewable fuel resources using MSWs.

Keywords: hydrothermal treatment, biochar production, biomass, municipal solid waste, renewable energy

1. Introduction

Recently, to address increasing energy consumption and the steady depletion of fossil fuel reserve, many investigations have been conducted to develop alternative and renewable energy resources using biomass and waste [1, 2]. Because rapid urbanization is occurring almost everywhere in the world, substantially increasing municipal solid waste (MSW), which is of great global concern. As an effective means for treating MSW, the mechanical biological treatment (MBT) system has received interest in Korea. MBT consists of two stages: mechanical treatment (MT) and biological treatment (BT). The MBT system was developed in Germany for waste processing [3]. In the first stage of this process (MT), MT is a size-based process that functions to remove all individual elements (regardless of size) that can be used for the production of refuse-derived fuel (RDF) (e.g., metal, plastic, paper, glass, and biodegradable materials). The larger MSW components are collected separately as combustible matter and then separated further to harvest materials for the production of refuse-derived fuel (RDF), which is used in energy generation. After the MT stage, the residual MSW is used for the production of organic fertilizers and biogas (CH_4) in the BT stage. However, the BT stage has many problems, such as a long treatment period requirement, (e.g., more than 1 week or 1 month) and the emission of unpleasant odors [1, 3, 4].

Thermo-treatment concept is the main conversion technology to develop the BT stage, such as carbonization (400–500°C), pyrolysis (500–600°C), gasification (600–1000°C), and combustion (800–1000°C) to produce carbon-neutral energy from several kinds of biomass wastes. Hydrothermal treatment, which is a thermo-chemical conversion process that employs subcritical water (water heated to any temperature less than its critical temperature of 373°C under sufficient pressure to maintain the liquid state), functions by hydrolyzing biomass components that contribute greatly to the decomposition of structural biomass compound, the major constituents of biomass contained in MSW [5–9]. Therefore, the properties and drying performance of biomass as an energy resource can be improved significantly in a short time [7, 10, 11].

In the current research, we employed a pilot-scale hydrothermal treatment system to generate alternative solid fuel products from MSW using subcritical water (200°C, 1.6 MPa). The MSW samples tested in this study were collected from an MSW treatment facility in Korea. The samples were at the MT stage of the MBT system and were mainly composed of food residue (40–50%) and paper (30–40%)—Chung et al. [12]. They had a high moisture content of ~50–60% due to the food residue. The physical and chemical characteristics of the MT residue needed to be altered (dehydrated, compacted, and upgraded) for it to be used as a solid fuel, such as RDF. Thus, cellulose, hemicellulose, and lignin were used as surrogate MSWs for food as well as paper wastes. Then, the effects of hydrothermal treatment on the conversion of the biomass comprising the MSW samples were examined by varying the reaction temperatures in the range of 150–280°C, and the changes in the biomass characteristics were investigated.

2. Materials and methods

2.1. Materials

Food residue and paper content, which comprised the highest proportions in the composition of the MT residue obtained from an MBT system in Mokpo city, Korea, were evaluated because the composition of the MSW varies according to each season (e.g., food waste and paper components can fluctuate between 70 and 80% of total MSW) [12], surrogate MSW (SM) residues, which were prepared using newspaper and Korean Kimchi instead of paper and food waste, were mixed at two different ratios (SM 1 and SM 2) after the crushing process: SM 1 = 5:5 and SM 2 = 3:7 (waste paper:Kimchi (wet, w/w)). **Table 1** shows the properties of these surrogate MSWs, including the results of proximate and ultimate analysis and the calorific values. In addition, pure cellulose (α-cellulose-fiberform, Nacalai Tesque Inc., Kyoto, Japan), xylan (Beechwood, SIGMA) which are the main components of hemicellulose [13], and lignin (Kanto Chemical Co., Inc., Japan) were also tested to investigate the effects of hydrothermal conversion on these materials.

	Paper	Kimchi	Raw surrogate MSW		Hydrothermally treated MSW	
			SM 1	SM 2	SM 1	SM 2
Moisture (a.r.)	2.3	92.4	47.4	65.4	66.3	79.6
Moisture[1]	–	–	36.8	54.2	12.6	17.3
Proximate analysis (wt.%, d.b.)						
Volatile matter	87.0	67.1	87.1	83.1	75.3	74.2
Fixed carbon	5.3	22.6	6.0	8.3	15.0	13.3
Ash	7.7	10.3	6.9	8.7	9.7	12.5
Ultimate analysis (wt.%, d.b.)						
C	40.3	33.6	37.0	35.6	43.5	41.7
H	5.6	5.3	5.4	5.4	5.7	5.3
N	0.2	3.5	1.9	2.6	1.1	0.4
O	46.4	47.3	48.8	47.7	40.0	40.1
Calorific value (MJ/kg, d.b.)	15.5	14.7	15.1	14.9	17.2	16.1

After natural drying for 24 h.
a.r., as received.
d.b., dry basis.

Table 1. Properties of raw and treated surrogate MSWs by hydrothermal treatment.

2.2. Pilot-scale hydrothermal treatment system

Hydrothermal treatment experiments were performed using a 200 L pilot-scale reactor (**Figure 1(a)**). The reactor consists of a steam boiler and a steam condenser. For all of the experiments, 20 kg of surrogate MSW was supplied to the reactor. The operating temperature of the hydrothermal treatment was set at 200°C, with a pressure of 1.6 MPa, and the reaction was carried out for 60 min. After the hydrothermal reaction was completed, the residual steam was discharged, and the products were collected from the reactor.

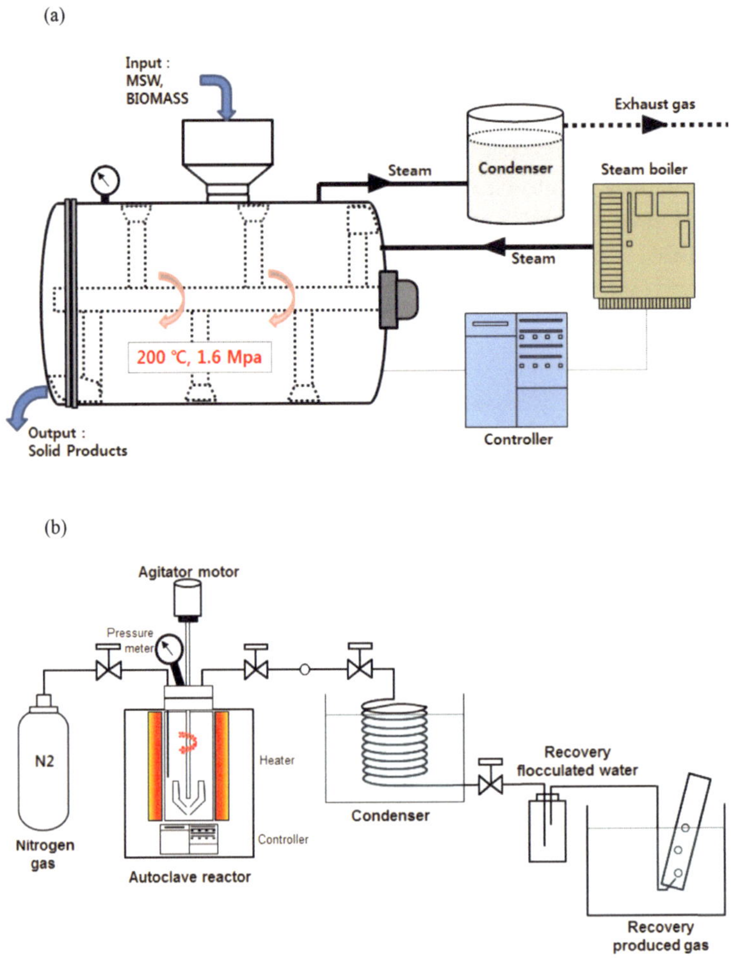

Figure 1. Schematics of (a) pilot- and (b) lab-scale hydrothermal treatment systems.

2.3. Lab-scale hydrothermal treatment reactor

A laboratory-scale hydrothermal treatment reactor was used to investigate the effects of hydrothermal treatment on the characteristic changes in pure cellulose, hemicellulose, and lignin. The experiments were performed using a 500 mL autoclave reactor (**Figure 1(b)**) consisting of a reactor body, heater, and steam condenser and operated under N_2 gas. A compound sample (20 g) was mixed with an equal amount of water and loaded into the reactor. The operating temperatures and pressures ranged from 150 to 280°C and 1.3 to 5.5 MPa, respectively, and the reaction time was 30 min. The components in the reactor were mixed vigorously using an agitator rotating at 200 rpm.

2.4. Analytical procedures

Elemental composition analysis of surrogate MSWs, cellulose, hemicellulose, lignin, and their solid products were carried out using a PerkinElmer 2400 Series II CHN organic elemental analyzer (PerkinElmer, Waltham, MA, USA). Proximate analysis was conducted using a SHIMADZU D-50 simultaneous TGA/DTA analyzer. Calorific values were determined using a bomb calorimetric standard method according to JIS M-8814 (JPN ISO1928:1995). The biomass compositions of the newspaper and Kimchi were analyzed by Nihon Hakko Shiryo Company (Kawasaki, Japan). The biomass compositions of the cellulose, hemicellulose, and lignin are defined in Eqs. (1)–(3) [14, 15].

$$\text{Hemicellulose}(\%) = \text{NDF} - \text{ADF} \tag{1}$$

$$\text{Cellulose}(\%) = \text{ADF} - \text{ADL} \tag{2}$$

$$\text{Lignin}(\%) = \text{ADL} \tag{3}$$

where NDF is neutral detergent fiber, ADF is acid detergent fiber, and ADL is acid detergent lignin.

3. Results and discussion

3.1. Properties of surrogate MSWs after hydrothermal treatment

Figure 2 shows the biochar products from MSW. Furthermore, **Table 1** and **Figure 3(a)** show the properties of raw and treated SM 1 and SM 2 using the pilot-scale hydrothermal treatment reactor operated at 200°C.

Figure 2. Biochar products process from waste by hydrothermal carbonization; (a) surrogated municipal solid waste, (b) pilot-scale hydrothermal treatment systems, and (c) biochar products.

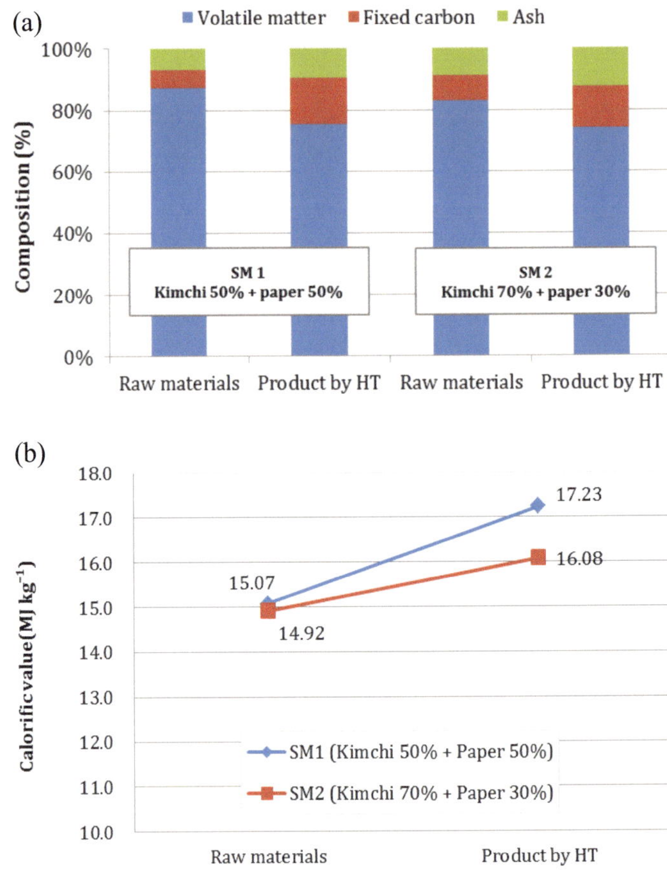

Figure 3. Enhancement of properties of surrogate MSWs by hydrothermal treatment; (a) the proximate analysis and (b) the calorific values.

The moisture content in the raw SM 1 and SM 2 samples, decreased significantly after 24 h of natural drying. However, the moisture content of the treated SM 1 and SM 2 samples decreased from 66.3 to 12.6% and 79.6 to 17.3%, respectively, after 24 h of natural drying. Hydrothermal treatment is known to dissociate physical and chemical structures of biomass in surrogate MSWs [5, 16]; therefore, the biomass in MSWs is decomposed into small and simple molecules. Our results demonstrated that hydrothermal treatment improved the drying performance of the surrogate MSWs. Additionally, the chemical properties of the surrogate MSWs were changed using the hydrothermal treatment. The raw SM 1 and SM 2 samples contained high volatile matter (87.1 and 83.1%, respectively) and oxygen (48.8 and 47.7%, respectively) contents, which are similar to the volatile matter and oxygen contents of other biomass materials [13, 17, 18]. After the hydrothermal treatment, the volatile matter and oxygen contents decreased, whereas the fixed carbon content increased from 6.0 to 15.0% for SM 1 and from 8.9 to 13.3% for SM 2 via hydrolysis reactions (**Figure 3a** and **b**) shows the calorific values of the surrogate MSWs (SM 1 and SM 2) before and after the hydrothermal treatment. After the hydrothermal treatment, the calorific values of SM 1 and SM 2 samples increased from 15.1 to 17.2 MJ/kg and 14.9 to 16.1 MJ/kg, respectively, which resulted in an increase of fixed carbon content. The increase in the calorific value of SM 1 was greater than that of SM 2 due to the higher paper content in the MSWs. This result suggests that the hydrothermal treatment is more effective at producing an upgraded solid fuel from MSW with high paper content.

3.2. Effect of hydrothermal treatment on upgrading of biomass in MSW

The surrogate MSW was prepared by blending newspaper and Kimchi. The content of the three major biomass components (cellulose, hemicellulose, and lignin) in the MSW samples was considered to be an important factor affecting the performance of the hydrothermal treatment. The characteristic changes of these components were analyzed. Additionally, the effects of hydrothermal treatment on the calorific values of cellulose, hemicellulose, and lignin were also examined.

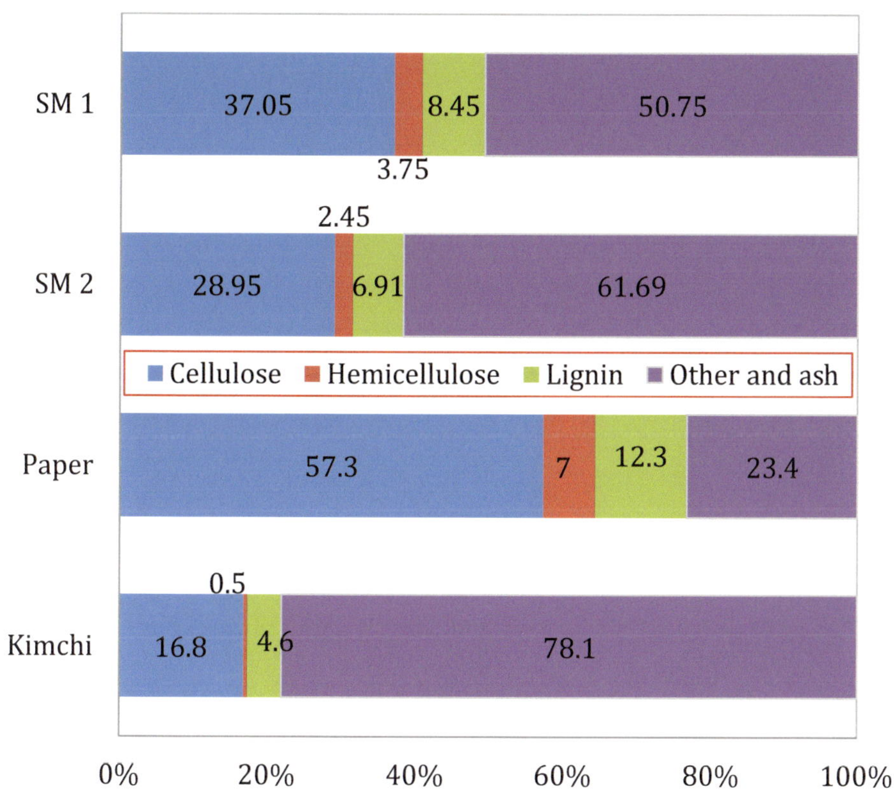

Figure 4. Biomass composition in the surrogate MSWs and their comparison with those of newspaper and Kimchi.

3.2.1. Biomass composition of MSW

Figure 4 shows the composition of the biomass components comprising the surrogate MSW. Dried newspaper was composed of 57.2 cellulose, 12.3 lignin, and 6.9% hemicellulose. Dried Kimchi was composed of 16.8 cellulose, 4.6 lignin, and 0.5% hemicellulose. Therefore, the total cellulose, lignin, and hemicellulose content of the newspaper and Kimchi were 76.5 and 29.1%, respectively. The total cellulose, lignin, and hemicellulose content in the biomass of the SM 1 and SM 2 samples were 49.4 and 38.4%, respectively. Specifically, the SM 1 and SM 2 samples contained 37.1 and 29.0% cellulose, 8.5 and 6.9% lignin, and 3.8 and 2.5% hemicellulose, respectively (**Figure 4**). The composition of biomass components (cellulose, hemicellulose, and lignin) of the MSW was found to influence the properties of the hydrothermal products.

3.2.2. Changes in the properties of biomass components

Hydrothermal treatment changed the properties of cellulose, hemicellulose, and lignin.

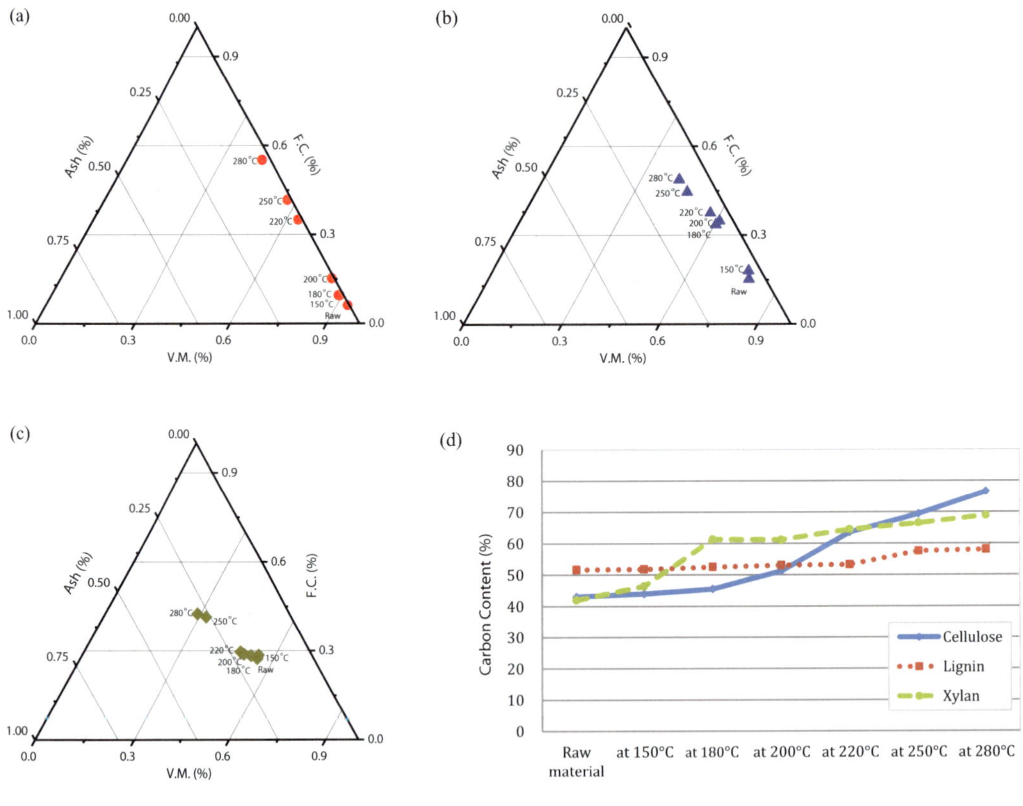

Figure 5. The changing characteristics of biomass components by hydrothermal treatment; (a) cellulose, (b) hemicellulose, (c) lignin, and (d) carbon content.

Cellulose							
	Raw cellulose	At 150°C	At 180°C	At 200°C	At 220°C	At 250°C	At 280°C
Proximate analysis (wt.%, d.b.)							
Volatile matter (d.b.)	93.4	88.9	89.3	84.0	63.7	56.9	42.2
Fixed carbon (d.b.)	6.1	9.6	9.5	15.2	35.0	41.7	55.1
Ash (d.b.)	0.5	1.6	1.3	0.8	1.4	1.4	2.7
Ultimate analysis (wt.%, d.b.)							
C	43.0	43.9	45.5	51.2	63.5	69.4	76.5
H	6.4	6.5	6.0	5.7	4.7	4.6	4.5
N	0.0	0.0	0.0	0.0	0.0	0.0	0.0
O	50.1	48.0	47.2	42.3	30.4	24.6	16.3
Calorific value (MJ/kg, d.b.)	16.5	16.6	18.9	23.0	26.5	26.8	27.7
Hemicellulose							
	Raw xylan	at 150°C	at 180°C	at 200°C	at 220°C	at 250°C	at 280°C
Proximate analysis (wt.%, d.b.)							
Volatile matter (d.b.)	79.8	78.3	60.5	61.0	56.8	46.1	41.6
Fixed carbon (d.b.)	15.2	17.9	33.5	34.8	37.5	44.6	48.7
Ash (d.b.)	5.1	3.8	6.0	4.3	5.7	9.4	9.7
Ultimate analysis (wt.%, d.b.)							
C	41.9	46.4	61.3	61.2	64.5	66.5	68.9
H	6.0	5.3	5.2	5.0	5.0	5.0	4.9
N	0.0	0.0	0.0	0.0	0.0	0.0	0.0
O	47.0	44.5	27.5	29.5	24.8	19.1	16.5
Calorific value (MJ/kg, d.b.)	13.9	14.9	21.5	22.6	24.9	25.6	26.5
Lignin							
	Raw lignin	at 150°C	at 180°C	at 200°C	at 220°C	at 250°C	at 280°C
Proximate analysis (wt.%, d.b.)							
Volatile matter (d.b.)	54.8	54.6	52.4	50.1	48.7	32.2	28.9
Fixed carbon (d.b.)	27.4	28.6	28.4	28.8	29.6	41.5	42.5

Cellulose							
Ash (d.b.)	17.8	16.8	19.2	21.1	21.7	26.3	28.6
Ultimate analysis (wt.%, d.b.)							
C	51.6	51.8	52.5	53.0	53.2	57.5	58.0
H	4.3	4.4	4.2	4.1	3.9	3.5	3.4
N	0.0	0.0	0.0	0.0	0.0	0.0	0.0
O	26.3	26.0	24.1	21.8	21.2	12.7	10.0
Calorific value (MJ/kg, d.b.)	20.4	20.1	21.8	22.8	23.1	25.2	26.0

d.b., dry basis.

Table 2. Effect of hydrothermal treatment on changes in the properties of cellulose, hemicellulose, and lignin.

Along with the results of the ultimate analysis, **Figure 5** and **Table 2** show the results of the proximate analysis of the biomass components by varying the hydrothermal reaction temperature. The fixed carbon content of cellulose increased from 6.1 to 35.0% in response to hydrothermal treatment at 220°C (**Table 2** and **Figure 5(a)**). This result suggests that the cellulose began to decompose at 220°C. When xylan was used as a hemicellulose, the fixed carbon and carbon contents increased from 15.2 to 33.5% and from 41.9 to 61.3%, respectively, at 180°C (**Figure 5(b)**). Below 180°C, the compositions of these products were not different from those of the raw material. This result is not surprising because the hydrolysis of hemicellulose occurs at 180°C [6, 11]. As the fixed carbon content and carbon contents increased due to the hydrothermal treatment, the calorific value increased. However, the results for lignin were different from those for cellulose and hemicellulose. Lignin started to decompose at temperatures exceeding 250°C (**Figure 5(c)**). This can most likely be attributed to the decomposition or pyrolysis of lignin at temperatures slightly below 250°C.

As a result, the increase in the fixed carbon content of the surrogate MSWs (SM 1 and SM 2) was influenced by the increase in the fixed carbon content of cellulose and hemicellulose (**Figure 5(d)**). Additionally, the ash content of lignin was higher than that of cellulose, indicating that lignin possesses higher ash content.

3.2.3. *Changes in calorific value and energy recovery efficiency*

Figure 6(a) shows the calorific values of cellulose, hemicellulose, and lignin. Pure lignin had a higher calorific value (20.4 MJ/kg) than pure cellulose (16.5 MJ/kg) and hemicellulose (13.9 MJ/kg). When cellulose was treated by hydrothermal treatment, the calorific values increased to 18.9, 23.1, 26.5, and 27.7 MJ/kg at 180, 200, 220, and 280°C, respectively. When xylan was used as a hemicellulose, the calorific value increased from 13.9 to 21.5, 22.6, 25.0, and 26.5 MJ/kg at 180, 200, 220, and 280°C, respectively. However, the calorific value of lignin started to increase from 20.4 to 25.2 MJ/kg at approximately 250°C. The increase in the calorific values for SM 1 and SM 2 was influenced by their cellulose and hemicellulose content. As reaction temperatures increase, the calorific value of solid products increases, but the amount of solid product decreases due to

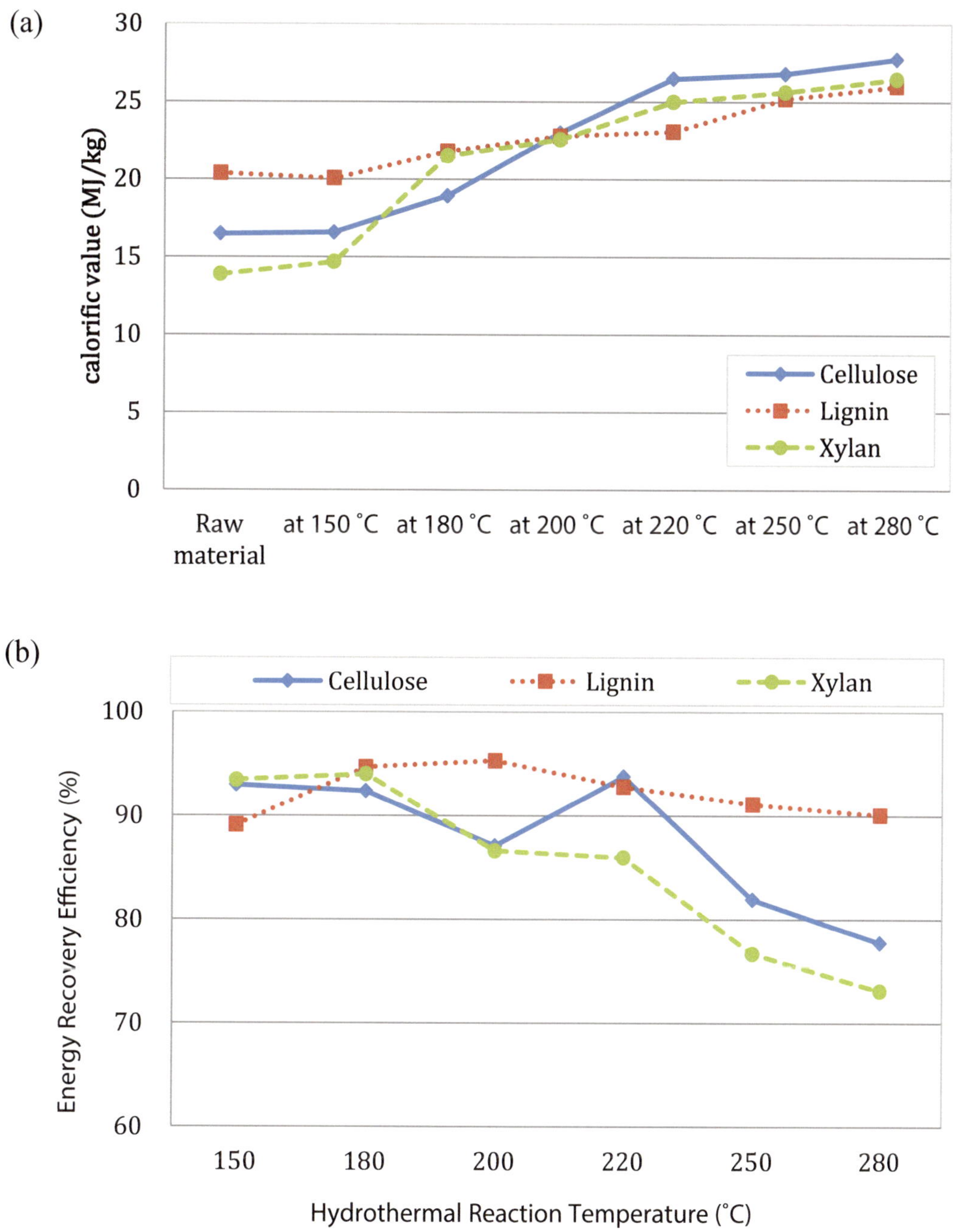

Figure 6. Alternation of properties in biomass components by hydrothermal treatment; (a) the calorific values and (b) energy recovery efficiency (ERE).

chemical dehydration and decarboxylation (i.e., removal of CO_2) [7, 18–21]. Therefore, in hydrothermal reactions, an optimum temperature should be maintained to maximize the energy recovery efficiency (ERE) [22–24]. The ERE is an important parameter for evaluating the effect of hydrothermal treatment on solid fuel production (**Figure 6(b)**). The ERE is defined by Eq. (4):

$$\text{ERE} = (\text{calorific value of solid product} \times \text{mass of dried solid product}) / (\text{calorific value of raw materials} \times \text{mass of dried raw material}) \quad (4)$$

The highest ERE of each material indicates the optimum reaction temperature for hydrothermal treatment. Cellulose decomposition began at 220°C, and the optimum reaction temperature used to maximize the ERE was 220°C (ERE = 93.8%). When xylan was used as a hemicellulose, it began to decompose at 180°C, and the optimum reaction temperature was 180°C (ERE = 94.0%). Lignin began to decompose at 250°C, and the maximum ERE was 95.3% at a reaction temperature of 200°C.

These results suggest that the optimum temperature of hydrothermal treatment to produce more energy-rich solid fuel is approximately 200°C. Fundamentally, pure lignin has a high calorific value (20.4 MJ/kg) and constitutes approximately 4–12% of solid fuel sources. When treated at 200°C, the calorific values of cellulose and hemicellulose were improved and the total remaining solid mounts were increased. Consequently, the optimum ERE of the MSWs and biomass products after hydrothermal treatment was obtained.

3.3. Mechanism of hydrothermal treatment for upgrading solid products

The results indicated that the hydrothermal treatment induce hydrolysis, chemical dehydration, and decarboxylation reactions (**Figure 7**). Hydrolytic reactions caused by 1 mol of water cleave cellulose and hemicellulose at ester and ether bonds of cellulose and hemicellulose into smaller molecules [6, 7, 11, 25]. This hydrolysis reaction can complete the conversion of biomass within a few reaction cycles. Furthermore, the fuel properties of the products generated

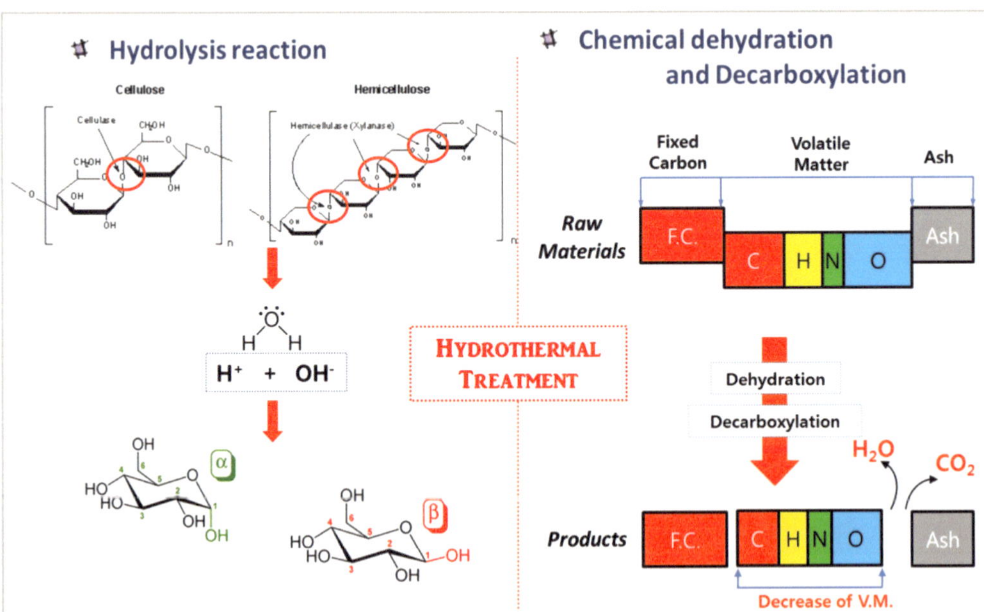

Figure 7. Mechanism of hydrothermal treatment to upgrade solid products.

from biomass (i.e., cellulose, hemicellulose, and lignin) have been shown to be upgraded via chemical dehydration and decarboxylation reactions that released H_2O and CO_2 [6, 7, 11, 26]. Along with a loss in weight, these reactions caused a decrease in volatile matter and an increase in carbon content in the biomass products compared with the raw materials. These effects can be utilized for the drying and carbonization of biomass into an alternative fuel.

3.4. Coalification band of hydrothermal products

Hydrothermal treatment can upgrade the properties of the biomass components of MSWs in a manner similar to the coalification process. The coalification bands of raw and treated surrogate MSWs (SM 1 and SM 2) were compared with the coalification band of pure cellulose, hemicellulose, lignin, and various types of coal (**Figure 8**). MSW is known to have high H/C and O/C ratios, which is similar to those of cellulose and other biomass materials [6, 17, 18, 21]. The H/C and O/C ratios of SM 1, SM 2, cellulose, hemicellulose, and lignin decreased with the coalification status between lignite and sub-bituminous coal. This occurs when the biomass components of MSW are converted into carbonaceous products by chemical dehydration reactions during hydrothermal treatment ($4(C_6H_{10}O_5)n \Leftrightarrow 2(C_{12}H_{10}O_5)n + 10H_2O$) [6, 10, 11, 18, 20]. Due to dewatering, dehydration, and decarboxylation, hydrothermal reactions can enhance biomass properties by reducing the hydrogen and oxygen contents of reaction products, resulting in increased calorific values of biomass products.

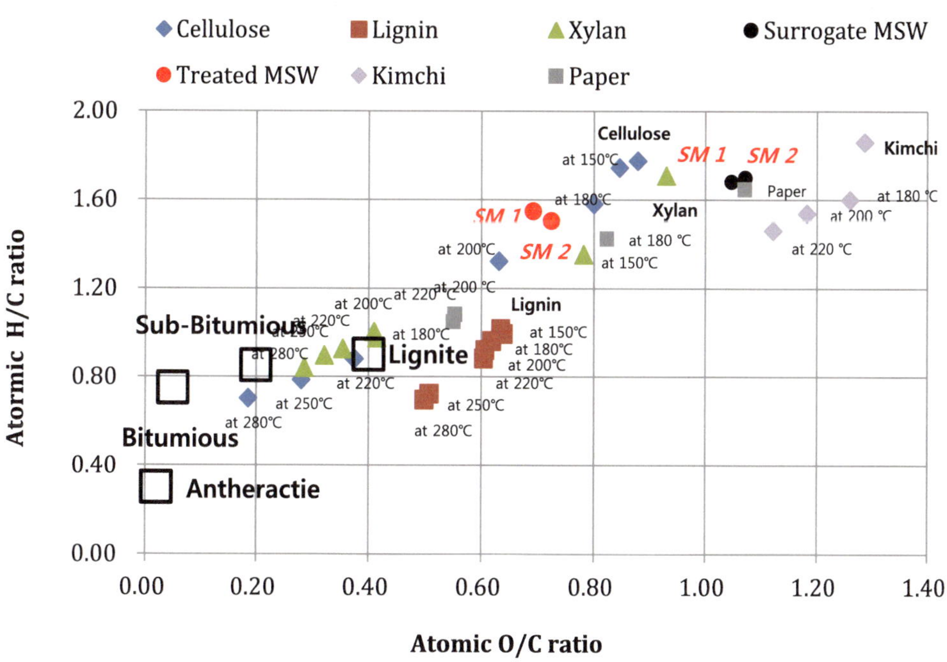

Figure 8. Comparison of coalification bands of raw and treated surrogate MSWs.

4. Conclusion

The effects of hydrothermal treatment on the properties of MSWs were investigated for their conversion into fuel products with high-energy efficiencies. After the hydrothermal treatment, the surrogate MSWs containing high paper content demonstrated significant increases in their carbon content and calorific values. Therefore, cellulose, hemicellulose, and lignin that constitute the MSWs as biomass were used to investigate the effects of the reaction temperature. The optimum reaction temperature for a mixture of cellulose, hemicellulose, and lignin was found to be approximately 200°C. As a result, the status of the treated products corresponded with solid fuels between lignite and sub-bituminous coal.

Author details

Daegi Kim[1]*, Ki Young Park[2] and Kunio Yoshikawa[3]

*Address all correspondence to: daegi.kim81@gmail.com

1 Bio Resource Team, Plant Engineering Division, Institute for Advanced Engineering, Beagam-myeon, Cheoin-gu, Yongin-si, Gyeonggi-do, Korea

2 Department of Civil and Environmental Systems Engineering, Konkuk University, Gwangjin-gu, Seoul, Korea

3 Department of Environmental Science and Technology, Tokyo Institute of Technology, Nagatsuta-cho, Midori-ku, Yokohama, Japan

References

[1] Barrena R, Imporzano G, Ponsa S, Gea T, Artola A, Vazquez F, Sanchez A, Adani F. In search of a reliable technique for the determination of the biological stability of the organic matter in the mechanical-biological treated waste. Journal of Hazardous Materials. 2009;**162**:1062-1072. DOI: 10.1016/j.jhazmat.2008.05.141

[2] Siddharth J, Sharma MP. Power generation from MSW of Haridwar: A feasibility study. Renewable and Sustainable Energy Reviews. 2011;**15**:66-90. DOI: 10.1016/j.rser.2010.09.007

[3] Velis CA, Longhurst PJ, Drew GH, Smith R, Pollard SJ. Production and quality assurance of solid recovered fuels using mechanical—biological treatment (MBT) of waste: a comprehensive assessment. Critical Reviews in Environmental Science and Technology. 2010; **40**(12): 979-1105. DOI: 10.1080/10643380802586980

[4] Yun YS, Park JI, Park JM. High-rate slurry-phase decomposition of food waste: Indirect performance estimation from dissolved oxygen. Process Biochemistry. 2005;**40**:1301-1306. DOI: 10.1016/j.procbio.2004.05.003

[5] Byrappa K, Adschiri T. Hydrothermal technology for nanotechnology. Progress in Crystal Growth and Characterization of Materials. 2007;**53**:117-166. DOI: 10.1016/j.pcrysgrow.2007.04.001

[6] Fenke A, Zidgler F. Hydrothermal carbonization of biomass: A summary and discussion of chemical mechanisms for process engineering. Biofuels, Bioproduct and Biorefining. 2010;**4**:160-177. DOI: 10.1002/bbb.198

[7] He W, Li G, Kong L, Wang H, Huang J, Xu J. Application of hydrothermal reaction in resource recovery of organic wastes. Resources, Conservation and Recycling. 2008;**52**:691-699. DOI: 10.1016/j.resconrec.2007.11.003

[8] Pandji P, Namioka T, Yoshikawa K. Coal alternative fuel production from municipal solid wastes employing hydrothermal treatment. Applied Energy. 2012;**90**:298-304. DOI: 10.1016/j.apenergy.2011.03.021

[9] Weber Walter J, Han Kim S. Removal of greases and lubricating oils from metal parts of machinery processes by subcritical water treatment. The Journal of Supercritical Fluids. 2013;**80**:30-37. DOI: 10.1016/j.supflu.2013.03.034

[10] Hwang IH, Aoyama H, Matsuto T, Nakagishi T, Matsuo T. Recovery of solid fuel from municipal solid waste by hydrothermal treatment using subcritical water. Waste Management. 2012;**32**:410-416. DOI: 10.1016/j.wasman.2011.10.006

[11] Kim D, Pandji P, Yoshikawa K. Hydrothermal upgrading of Korean MSW for solid fuel production: Effect of MSW composition. Journal of Combustion. 2012;2012:8 p. DOI: 10.1155/2012/781659

[12] Chung IH, Kim D, Chae JH. The study on possibility of RDF by analysis of physico-chemical characteristics of MSW by the size distribution. Journal of Korean Society of Waste Management. 2009:**26**(7);626-634

[13] Laine, C. Structures of Hemicelluloses and Pectins in Wood and Pulp [Dissertation for the degree of Doctor]. Helsinki University of Technology, Espoo, Finland; 2005

[14] Carriera M, Loppinet-Serania A, Denuxa D, Lasnierc J, Ham-Pichavantc F, Cansella F, Aymoniera C. Thermogravimetric analysis as a new method to determine the lignocellulosic composition of biomass. Biomass and Bioenergy. 2011:**35**;298-307. DOI: 10.1016/j.biombioe.2010.08.067

[15] Verardi A, Ricca E, De Bari I, Calabrò V. Hydrolysis of Lignocellulosic: Biomass Current Status of Processes and Technologies and Future Perspectives. Intech Open Access Publisher, Rende (CS), Italy; 2012.

[16] Kim D, Yoshikawa K, Lee K, Park KY. Investigation of the combustion characteristics of municipal solid wastes and their hydrothermally treated products via thermogravimetric analysis. Journal of Material Cycles and Waste Management. 2015:**17**(2);258-265. DOI: 10.1007/s10163-014-0342-y

[17] Tri Mursito A, Hirajima T, Sasaki K, Kumagai S. The effect of hydrothermal dewatering of Pontianak tropical peat on organics in wastewater and gaseous products. Fuel. 2010:**89**;3394-3924. DOI: 10.1016/j.fuel.2010.06.035

[18] Yuliansyah AT, Hirajima T, Kumagai S, Sasaki K. Production of solid biofuel from agriculture waste of the Palm oil industry by hydrothermal treatment. Waste and Biomass Valorization. 2010:**1**(4);395-405. DOI: 10.1007/s12649-010-9045-3

[19] Simsek kus N. Organic reactions in subcritical and supercritical water. Tetrahedron. 2012:**68**;949-958. DOI: 10.1016/j.tet.2011.10.070

[20] Tri Mursito A, Hirajima T, Sasaki K. Upgraging and dewatering of raw tropical peat by hydrothermal treatment. Fuel. 2010:**89**;635-641. DOI: 10.1016/j.fuel.2009.07.004

[21] Xiao LP, Shi ZJ, Xu F, Sun RC: Hydrothermal carbonization of lignocellulose biomass. Bioresource Technology. 2010:**118**;619-623. DOI: 10.1016/j.biortech.2012.05.060

[22] Berge ND, Ro KS, Mao J, Flora JRV, Chappell MA, Bae S: Hydrothermal carbonization of municipal waste streams. Environmental Science and Technology 2011:**45**(13);5696-5703. DOI: 10.1021/es2004528

[23] Daegi K, Kwanyong L, Park KY. Hydrothermal carbonization of anaerobically digested sludge for solid fuel production and energy recovery. Fuel. 2014:**130**;120-125. DOI: 10.1016/j.fuel.2014.04.030

[24] Parshetti Ganesh K, Kent Hoekman S, Balasubramanian R. Chemical, structural and combustion characteristics of carbonaceous products obtained by hydrothermal carbonization of palm empty fruit bunches. Bioresource Technology. 2013:**135**;683-689. DOI: 10.1016/j.biortech.2012.09.042

[25] Quitain AT, Faisal M, Kang K, Daimon H, Fujie K. Low-molecular-weight carboxylic acids produced from hydrothermal treatment of organic wastes. Journal of Hazardous Materials. 2002:**B93**;209-220. DOI: 10.1016/S0304-3894(02)00024-9

[26] Jenkins BM, Baxter LL, Miles Jr TR, Mines TR. Combustion properties of biomass. Fuel Processing Technology. 1998:**54**;17-46. DOI: 10.1016/S0378-3820(97)00059-3

Section 4

Applications of Biochar

Chapter 4

Effects of Biochar on Plant Growth and Cadmium Uptake: Case Studies on Asian Lotus (*Nelumbo nucifera*) and Chinese Sage (*Salvia miltiorrhiza*)

Daike Tian, Amei Liu and Yanci Xiang

Additional information is available at the end of the chapter

http://dx.doi.org/10.5772/intechopen.68251

Abstract

Application of biochar has many benefits in agriculture, to understand benefits of biochar in crop production and remediation of heavy metal pollution, Asian lotus (*Nelumbo nucifera*) as an aquatic crop and Chinese sage (*Salvia miltiorrhiza*) as a traditional medicinal herb were used to evaluate biochar's effects on plant growth and cadmium (Cd) accumulation in plants in the artificially Cd-polluted condition in containers. In both cases, adding biochar (4% to 32% in soil mix) significantly increased plant biomass. However, its impact on plant physiological traits were unclear. In Asian lotus, the Cd content in rhizomes, petioles, and leaves significantly increased by 69%, 81% and 55%, respectively as 32% biochar added. Meanwhile, a maximum reduction (71%) showed on bioaccumulation coefficient of Cd, and an up to 1.3 folds increase occurred on Cd transfer coefficient of underground to aboveground tissues, which indicated that biochar effectively prevented Cd uptake in major edible parts. In Chinese sage, adding 32% biochar significantly decreased Cd content in leaves and roots by 52.81% and 43.63%, respectively. Therefore, as a valuable soil amendment of improving plant growth and reducing heavy metal uptake, biochar has a huge potential in green agriculture production and remediation of heavy-metal polluted environment.

Keywords: biochar, heavy metal pollution, cadmium, *Nelumbo nucifera*, lotus, *Salvia miltiorrhiza*, Chinese sage, danshen, plant growth, vegetable, medicinal plant

1. Introduction

With the rapid development of urbanization, industry, and agriculture, heavy metal contamination in water, soil, and the atmosphere has been increasing. Heavy metals accumulate in plants, animals, and human bodies, therefore, human health is exposed to a great potential threat through the food chain. Biochar is an emerging environment functional material and has been proven to be an efficient adsorbent for toxic organic pollutants and heavy metals [1–4]. With excellent physicochemical structure, biochar has great agricultural benefits and the potential for remediation of contaminated soils by improving soil physicochemical properties [5, 6], increasing microbial activity in soil [7, 8], reducing nutrient leaching [9], and enhancing nutrient availability [10], as well as directly or indirectly decreasing the contaminant bioavailability in water and soil. Additionally, biochar is a renewable material that is low cost, sustainable, and environment friendly. It could be an effective way of utilization of agricultural and forestry wastes and has a wide application in sustainable agriculture. Previous studies [1, 11] have shown that biochar improved plant growth and inhibited migration and phytotoxicity of heavy metals. But, it is yet unknown on how to apply biochar in polluted water and how much the accumulation of cadmium (Cd) could be reduced in major aquatic crops such as Asian lotus (*Nelumbo nucifera* Gaertn.). In addition, no data is available about the optimum amount of biochar applied in polluted environment in order to produce safe food and medicine from aquatic plants.

With increasing human health issues caused by heavy metal contamination, a strict standard on the maximum of heavy metal content has been widely established in food and medicinal products. The upper limit set by European Commission in Regulation No. 1881/2006 was at 0.10 and 0.20 mg/kg wet weight in stem vegetables and fresh herbs, respectively. The Chinese Ministry of Health released GB 2762-2012 (Limiting Chemical Contaminants in Food) with a Cd cutoff value of 0.1 and 0.2 mg/kg in root and leafy vegetables, respectively. Cd is one of the top five toxic heavy metals with strong chemical activity, high toxicity, and difficulty to be degraded. Its contamination in major crops such as rice has become a severe problem in China in recent years [12, 13], and excessive Cd content has limited the export of crops and Chinese medicines. Therefore, it is of great scientific value and practical significance to investigate food safety and pollution remediation in heavy metal-contaminated environment.

Asian lotus is widely grown in China, Japan, and other Asian countries as an important vegetable, medicinal, and ornamental plant. Nearly, all parts of lotus plant are edible or used for medicinal purpose [14]. Lotus deserves attention because of its economic potential and because its product quality and economic value are influenced by heavy metal pollution [15, 16]. *Salvia miltiorrhiza* Bunge (danshen in China, red sage or Chinese sage in other areas) has been widely used as one of the most important traditional Chinese herbs in China, Korea, Japan, and other Asian countries [17]. Its dried roots contain major active constituents including various kinds of tanshinones that pharmacologically function as treatment for ischemic cardiovascular diseases, microcirculatory disturbance-related diseases, and coronary disorders [17, 18]. Cd has strong chemical activity and potentially high toxicity that creates challenges in remediation. Contamination of Cd is one of the most serious issues because it is a major biohazard in China.

To investigate effects of biochar on the growth of Asian lotus and Chinese sage, we conducted a trial in artificially Cd-polluted condition in containers to evaluate the toxicity and accumulation of Cd in both aquatic and terrestrial plants. The results of this study may provide a useful reference on restoration of heavy metal-polluted environment and production of safe and organic agricultural products.

2. Material and methods

2.1. Site details

The experiment was conducted in a plastic tunnel house and greenhouse, respectively, which are located at Shanghai Chenshan Plant Science Research Center (31°04′37″N; 121°10′35″E), Chinese Academy of Sciences, and Chenshan Botanical Garden, Shanghai of China. The light intensity is 7000 lx (2800–9300 lx) for both tunnel house and greenhouse during daylight hours. The temperature at 28°C (22–42°C) and humidity at 50% (30–70%) are for tunnel house, and the temperature at 25°C (20–36°C) and humidity at 60% (40–80%) are for greenhouse throughout the experiment.

2.2. Plant material

N. nucifera 'Taikong Lian 36', one of the most important lotus cultivars in China, and Chinese sage (*S. miltiorrhiza* Bunge.), a medicinal plant, were used for experimental plants. The lotus seeds were obtained from Guangchang White Lotus Development Bureau, Jiangxi of China, and those of *S. miltiorrhiza* were obtained from Northwest Agriculture and Forestry University, Shanxi, China.

2.3. Trial design and treatments

Pinewood biochar from a local market was sieved to use pieces about 0.5–1.0 cm in size (**Table 1**). Natural soil (loam) was air-dried (**Table 2**) and blended to create a uniform mix after removing rocks, twigs, leaf litters, etc. After recording dry weight, the soil was moistened to 40% maximum field water capacity with deionized water. The biochar was then added and thoroughly mixed with the soil at differing amounts: 0, 4, 8, 16, and 32% biochar-soil (v/v), and labeled as CK (control), B0, B4, B8, B16, and B32, respectively. The lotus seeds germinated in small containers, and the seedlings with three to four floating leaves were transplanted in larger pots without drainage holes for experimental treatment in early June, one plant per pot and six replication. Forty days later, except CK, each pot was added by the solution of 1.09 mg $Cd(NO_3)_2{\cdot}4H_2O$ (3 mg/kg Cd). The pots were placed under a transparent plastic tunnel house to avoid rainfall and regularly watered by tap water to retain the starting water level. No fertilizer was applied throughout trial. For a trial on Chinese sage, the soil-biochar mixture was made same as that in lotus experiment. One month later, the dissolved $Cd(NO_3)_2{\cdot}4H_2O$ (3 mg/kg Cd) was added to soil-biochar mixture other than CK. The lotus seedlings with three to four leaves were transplanted to the experimental pots with saucers in June, one seedling each pot and six pots per treatment. Pots were randomly arranged. Throughout experiment, plants were not fertilized but watered regularly with tap water to maintain soil moisture. To avoid possible leaching of heavy metal and other nutrients, the leachate from the pots was collected and later reapplied to the pots.

pH	Organic matter (cmol/kg)	Cation exchange capacity (cmol/kg)	Electrical conductivity (mS/cm)	Pore structure (%)	Cd content (mg/kg)
9.2	Immeasurable	23.16	0.69	Large 39%, small 44%	0.17

Table 1. Biochar properties.

Total nitrogen (g/kg)	Total phosphorus (g/kg)	Total potassium (g/kg)	pH	Organic matter (cmol/kg)	CEC (cmol/kg)	Electrical conductivity (mS/cm)	Cd content (mg/kg)
5.44	0.65	16.4	6.08	82.46	22.79	0.33	0.38

Table 2. Soil properties.

2.4. Data collection

2.4.1. Plant growth indicators

In lotus experiment, leaf number, leaf area, and plant height (equal to the height of the tallest leaf) were recorded at 35, 75, and 115 d, respectively, after transplanting. Leaf diameter and plant height were measured by a ruler. Six mature emergent leaves were randomly selected from each of the three pots to record leaf diameter ([long + short length diameter]/2), and then leaf area (πR^2) was calculated. Plant height was measured from the top of the pot to the top of the highest leaf in each pot. Seventy days after adding Cd, all the aboveground parts and underground parts (rhizomes only) were separately harvested. After soil removal, the rhizomes were immersed in a 0.2 μm EDTA solution for 30 min, followed by air-drying for 12 h, and then the fresh mass was determined. Thereafter, the weighed rhizomes were immediately placed into envelopes for drying at 65°C for 84 h in an oven (DHG-9240A, Shanghai Yiheng Scientific Instruments Co., Ltd., China), and then the dry mass was recorded.

In Chinese sage experiment, leaf number and area were recorded at 10, 35, and 60 d, respectively, after transplanting. The leaf area of both the largest and smallest mature leaves in each pot was measured by a leaf area meter (YMJ-A, Zhejiang Top Instrument Co., Ltd) to get the average value for each treatment (six plants). Eighty days after transplanting, leaves with stems and roots were harvested separately. After soil removal, root fresh weight was determined before immersing roots in a 0.2 μm EDTA solution for 30 min followed by air-drying for 4 h. Roots were then placed into envelopes and dried at 65°C for 48 h in a drying oven. Finally, the root dry weight was measured.

2.4.2. Plant physiological indicators

All plant samples were collected with a hole puncher (1 cm diameter) at the same position of each fresh leaf. Water extract of leaves was used to measure cell membrane permeability by tissue fluid exosmosis conductance method (n = 3/treatment) using a conductivity meter (DDS-11AT, Shanghai, China) [19]. The relative chlorophyll content of fresh leaves was recorded by a handheld chlorophyll meter SPAD-502 Plus (Konica Minolta Sensing, Inc., Japan; n = 6), according to Xiong [19]. Following Xiong [19], the thiobarbituric acid method (n = 3) was used to determine methane dicarboxylic aldehyde (MDA) content in fresh leaves and the NBT photochemical reduction method to analyze superoxide dismutase (SOD) activity. At the same time as the chromogenic reaction, enzyme-extracted leaf liquid was collected and tested at 560 nm OD by an ELISA plate reader (Tecan M200 PRO) (n = 3) [19].

2.4.3. *Cd content measurements*

Cd contents in the rhizomes, roots, leaves, and petioles were determined by Graphite Furnace Atomic Absorption Spectrometry, according to the "Determination of Cadmium in Foods GB/T 5009.15-2003" [20]. Microwave digestion method was carried out with a PE Analyst AA800 (Perkin Elmer) (n = 3/treatment).

The bioconcentration factor (BCF) reflects the degree or capability of accumulation of heavy metals from soil or water by plants. The biological transfer factor (TF) refers to the transference of heavy metals in plants in vivo Zayed et al. [21]. BCF and TF were calculated by BCF = trace element concentration in plant tissues at harvest/initial concentration of the element in the external nutrient solution and TF = heavy metal content ratio (Cd in aboveground tissues/Cd in rhizomes or roots).

2.5. Data analysis

The averages were determined by Microsoft Excel 2007. The analysis of variance (ANOVA), Duncan's multiple range test, and regression analysis were conducted by statistical package for social scientists (SPSS 16.0). Statistical significance was set at a level of $\alpha = 0.05$.

3. Results

3.1. Biochar promoted the growth of Asian lotus and Chinese sage

3.1.1. *Leaf growth*

In lotus experiment, no apparent leaf toxicity was observed 5 days after adding Cd. Thirty-five days later, the number, height, diameter, and leaf area of emerging leaves of *N. nucifera* 'Taikong Lian 36' reached the highest in the treatments with 32% biochar (**Table 3**). But, the relationship was unclear between the number of floating leaves and biochar ratio. Forty days later, the same plant indexes in treatment H0 were the lowest, only 32, 39, 59, and 29% of that in treatment H32, respectively. However, differences were not significant between H0 and the control. Therefore, Cd of low concentration could not cause strong toxicity. The relationships between biochar and the number, height, diameter, and leaf area of emerging leaves in lotus could be reflected by (1) $y = -0.03\,x^2 + 1.8\,x + 10.87$ ($R^2 = 0.955$, $P = 0.037$), (2) $y = -0.005\,x^2 + 1.042\,x + 27.047$ ($R^2 = 0.830$, $P = 0.170$), (3) $y = -0.002\,x^2 + 0.231\,x + 13.117$ ($R^2 = 0.836$, $P = 0.164$), and (4) $y = -0.000\,x^2 + 0.06\,x + 0.432$ ($R^2 = 0.962$, $P = 0.037$), respectively. It indicated that the growth of lotus was

Biochar rate (%)	Floating leaf number			Emerging leaf number			Emerging leaf height (cm)			Leaf diameter (cm)			Leaf area (m^2)		
	35 d with biochar	75 d with biochar and Cd	115 d with biochar and Cd	35 d with biochar	75 d with biochar and Cd	115 d with biochar and Cd	35 d with biochar	75 d with biochar and Cd	115 d with biochar and Cd	35 d with biochar	75 d with biochar and Cd	115 d with biochar and Cd	35 d with biochar	75 d with biochar and Cd	115 d with biochar and Cd
CK	8.2 ± 4.4 a	9.5 ± 5.5 a	11.5 ± 4.9 a	6.3 ± 3.7 b	7.5 ± 4.4 b	11.7 ± 6.2 b	8.4 ± 2.5 d	9.0 ± 5.0 c	21.8 ± 10.7 c	6.7 ± 0.7 d	9.2 ± 1.4 c	11.0 ± 1.4 c	0.1 ± 0.0 c	0.1 ± 0.1 c	0.2 ± 0.1 d
H0	7.20 ± 2.1 a	7.8 ± 2.5 a	10.3 ± 2.4 a	5.5 ± 3.4 b	5.8 ± 6.1 b	10.8 ± 11.7 b	8.5 ± 3.6 d	9.0 ± 6.6 c	22.3 ± 10.4 c	7.0 ± 1.1 d	9.4 ± 0.8 c	12.1 ± 2.5 c	0.1 ± 0.0 c	0.1 ± 0.1 c	0.4 ± 0.5 c
H4	7.7 ± 2.3 a	9.0 ± 2.7 a	11.3 ± 2.3 a	8.3 ± 2.6 b	9.8 ± 3.0 b	19.5 ± 6.1 b	16.7 ± 4.3 bc	21.2 ± 7.6 bc	39.5 ± 8.2 b	7.9 ± 0.9 cd	11.3 ± 1.4 bc	15.4 ± 3.1 b	0.1 ± 0.0 c	0.2 ± 0.1 c	0.9 ± 0.4 b
H8	10.5 ± 1.4 a	12.0 ± 3.6 a	15.0 ± 3.3 a	9.2 ± 3.1 b	11.5 ± 5.5 b	19.8 ± 11.9 b	14.5 ± 6.6 cd	17.0 ± 8.7 bc	33.0 ± 18.2 bc	8.4 ± 0.8 bc	11.5 ± 2.7 bc	15.1 ± 3.9 b	0.1 ± 0.1 c	0.3 ± 0.1 c	0.7 ± 0.3 b
H16	10.5 ± 1.6 a	12.5 ± 5.5 a	14.7 ± 5.1 a	15.5 ± 7.3 a	18.8 ± 9.6 a	32.8 ± 8.0 a	22.7 ± 6.4 b	25.3 ± 7.7 b	40.3 ± 12.1 b	9.3 ± 1.3 b	12.9 ± 1.4 b	15.5 ± 1.4 b	0.3 ± 0.1 b	0.5 ± 0.3 b	1.4 ± 0.5 a
H32	7.7 ± 4.3 a	8.0 ± 4.6 a	11.8 ± 3.2 a	20.0 ± 5.8 a	21.3 ± 6.1 a	34.3 ± 9.1 a	34.7 ± 8.4 a	39.0 ± 10.4 a	55.8 ± 5.4 a	11.2 ± 1.5 a	15.2 ± 2.6 a	18.7 ± 1.4 a	0.4 ± 0.2 a	0.7 ± 0.2 a	2.2 ± 0.6 a

CK refers to none of biochar and Cd added to the soil; H0, H4, H8, H16, and H32 refer to adding 0, 4, 8, 16, and 32% of biochar in soil, respectively. Different lower-case letters in the same column indicate significant difference ($P < 0.05$) among treatments.

Table 3. Effects of biochar on leaf growth of *N. nucifera* 'Taikong Lian 36' in Cd-polluted water.

significantly promoted by biochar added at 16–32% (with Cd added). The higher ratios of biochar promoted more leaf growth but also increased the cost (biochar cost).

In the experiment on Chinese sage, morphological leaf toxicity did not occur 10 d after adding Cd and biochar treatments. Both leaf number and area increased at 35 d after treatment and reached the highest value in the treatments with the highest biochar proportion (B32) at 60 d for the final measurement (**Table 4**). At 60 d, the leaf number of B4, B8, B16, and B32 increased 26.2, 30.9, 37.8, and 42.9%, respectively, while that of CK and B0 increased 19.4 and 26.0%, respectively, compared with those at 10 d. Likewise, leaf area of B4, B8, B16, and B32 increased 36.3, 50.9, 70.7, and 69.9%, respectively, while that of CK and B0 just increased 28.3 and 31.0%, respectively. Therefore, leaf growth of *S. miltiorrhiza* was significantly promoted with the addition of 16–32% biochar (with Cd added). Lower concentrations of Cd did not cause strong toxicity problems but resulted in the smaller leaf size and slower growth of plants.

3.1.2. Plant biomass

Biomass discrepancy is the most direct response to various environmental stimuli for plants. In lotus, our study showed that the total fresh weight, the fresh weight of the aboveground parts, and the fresh and dry weight of rhizomes increased by adding biochar (**Figure 1**). All the highest values of the tested plant growth indexes occurred in treatment H32. Among which, the highest total fresh weight was 4.5 times greater than H0. The fresh weight of underground parts made the greatest contribution to the total fresh weight, which correlated significantly with the proportion of biochar, and the maximum was 6.25 times greater than control values.

In Chinese sage, the total fresh weight, the leaf fresh weight, and the fresh and dry weight of roots increased with the addition of biochar in our study (**Figure 2**). All the highest values of the tested plant growth indices occurred in treatment B32 where the highest total fresh weight was 1.3 times of that in treatment B0. The fresh weight of roots made a greater contribution to

Biochar rate/%	Leaf number			Leaf area (cm^2)		
	10 d	35 d	60 d	10 d	35 d	60 d
CK	19.3 ± 3.2 b	21.8 ± 3.2 c	24.3 ± 4.0 b	37.8 ± 6.7 a	46.7 ± 4.00 ab	48.5 ± 5.6 bc
B0	24.5 ± 3.1 a	26.3 ± 2.9 ab	29.3 ± 7.3 ab	31.5 ± 4.3 a	39.8 ± 3.7 bc	41.3 ± 4.0 c
B4	21.0 ± 1.6 ab	23.5 ± 2.9 bc	26.5 ± 3.4 b	32.2 ± 4.8 a	38.6 ± 5.2 c	43.9 ± 6.4 bc
B8	23.5 ± 1.0 a	27.5 ± 2.1 a	30.8 ± 2.2 ab	33.9 ± 2.7 a	38.8 ± 7.2 c	51.2 ± 4.5 ab
B16	24.5 ± 1.9 a	30.0 ± 2.5 a	33.8 ± 1.5 a	35.1 ± 4.2 a	45.8 ± 4.2 bc	60.0 ± 3.8 a
B32	24.5 ± 1.0 a	29.3 ± 1.0 a	35.0 ± 2.7 a	34.3 ± 4.5 a	52.0 ± 1.2 a	58.1 ± 4.1 a

CK indicates no biochar and no Cd in the soil; B0, B4, B8, B16, and B32 indicate adding 0, 4, 8, 16, and 32% of biochar in soil, respectively. Different lower-case letters in the same column indicate significant differences ($P < 0.05$) among treatments.

Table 4. Effects of biochar on leaf growth of *S. miltiorrhiza* in Cd-polluted soil ($\bar{x} \pm s$).

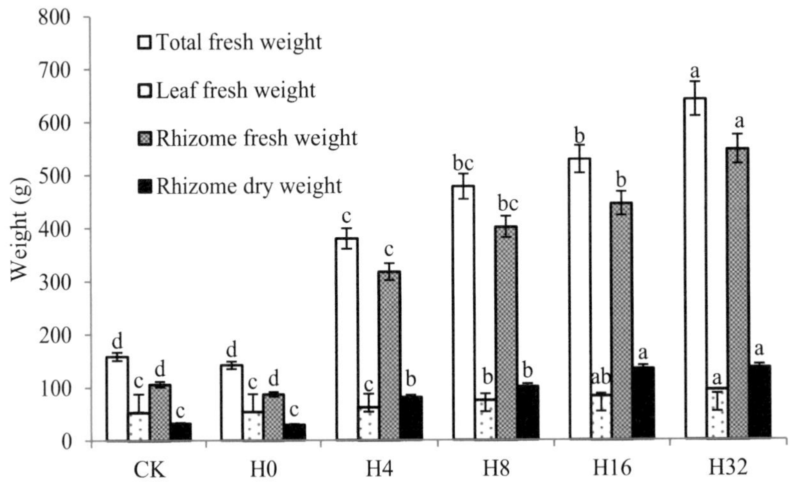

Figure 1. Effects of biochar on plant weight of *N. nucifera* 'Taikong Lian 36' in Cd-polluted water. CK refers to none of biochar and Cd added to the soil; H0, H4, H8, H16, and H32 refer to adding 0, 4, 8, 16, and 32% of biochar in soil, respectively. Different lower-case letters indicate significant difference ($P < 0.05$) among treatments.

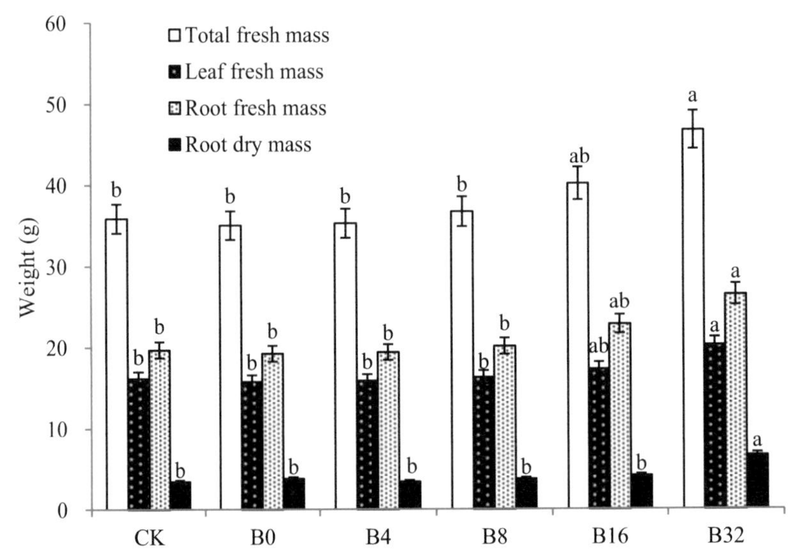

Figure 2. Effects of biochar on mass of *S. miltiorrhiza* in Cd-polluted soil. CK indicates no biochar and no Cd in soil; B0, B4, B8, B16, and B32 indicate adding 0, 4, 8, 16, and 32% of biochar in soil, respectively. Different lower-case letters indicate significant differences ($P < 0.05$) among treatments.

the total fresh weight than that of the leaves and stems, and it correlated significantly with the biochar proportion. The maximum root fresh and dry weight observed in B32 was 1.4 times of that in the control, which further verified that biochar promoted root growth.

3.2. Effects of biochar on physiological status

3.2.1. *Chlorophyll content*

The relative content of chlorophyll can be used to evaluate plant growth, physiological changes, and nitrogen nutrition. Three consecutive measurements on leaf chlorophyll were taken by SPAD at 20, 50, and 80 d after adding biochar (**Table 5**). The relative chlorophyll contents in leaves of treated plants were significantly higher than those in the control, but no significant difference was found between biochar treatments. The highest values in each treatment occurred at 10 d after adding Cd when the leaf growth reached its peak. The highest chlorophyll content (82%) in H0, followed by 62% in H16, could be explained by that Cd did not reach a toxic stress concentration affecting the enzyme activity relevant to chlorophyll formation. On the contrary, low concentration of Cd promoted chlorophyll synthesis. Biochar increased the relative chlorophyll content of lotus leaf. The chlorophyll formation was significantly promoted at 20 and 50 d after biochar being added. Fifty days later, it decreased slowly as plants started senescence.

In Chinese sage, the relative chlorophyll content was tested at 10, 35, and 60 d, respectively, after transplanting (**Table 6**). Relative chlorophyll contents in the leaves of treated plants were higher than those in the control, but no difference was found between the treatments of biochar ratios. The highest chlorophyll content in all treatments occurred at 35 d when the leaves were at the peak stage of growth. The highest increase rate of chlorophyll content from 10 to 35 d was in treatment B32 (42%), indicating that Cd might affect the enzyme activity relevant to chlorophyll synthesis, and it stimulated the chlorophyll formation at a certain concentration. Sixty days after biochar treatment, the relative chlorophyll content began to decrease slowly as the plants gradually senesced.

3.2.2. *Cell membrane permeability*

In both lotus and Chinese sage, the maximum of cell membrane conductivity occurred in treatment H0 and was significantly different from the control and biochar treatments (**Figure 3A** and **B**). This

Biochar rate (%)	Relative chlorophyll content (20 d with biochar)	Relative chlorophyll content (50 d with biochar)	Relative chlorophyll content (80 d with biochar)
CK	12.7 ± 0.8 ab	20.0 ± 2.1ab	15.8 ± 0.5c
H0	11.8 ± 2.0a	21.4 ± 0.9b	16.0 ± 1.4abc
H4	14.2 ± 0.9ab	22.0 ± 1.6a	14.9 ± 0.7 ab
H8	13.7 ± 0.8 ab	21.6 ± 1.3a	15.9 ± 0.8 abc
H16	14.1 ± 0.8 ab	22.7 ± 0.5 a	15.3 ± 0.9 a
H32	13.1 ± 1.8b	20.8 ± 1.6ab	14.8 ± 0.27 bc

CK refers to none of biochar and Cd added to the soil; H0, H4, H8, H16, and H32 refer to adding 0, 4, 8, 16 and 32% of biochar in soil, respectively. Different lower-case letters in the same column indicate significant difference ($P < 0.05$) among treatments.

Table 5. Effects of biochar on relative chlorophyll content of *N. nucifera* 'Taikong Lian 36' in Cd-polluted water.

Biochar rate (%)	Relative chlorophyll content		
	10 d	35 d	60 d
CK	16.9 ± 0.2 a	22.1 ± 0.4 a	14.5 ± 0.4 ab
B0	16.6 ± 0.8 ab	22.5 ± 0.8 a	13.0 ± 0.5 ab
B4	16.0 ± 1.55 b	22.1 ± 0.7 a	12.4 ± 0.9 bc
B8	16.5 ± 0.8 ab	22.9 ± 1.2 a	12.0 ± 0.9 c
B16	16.2 ± 0.5 ab	22.3 ± 1.3 a	12.3 ± 0.9 bc
B32	16.2 ± 0.5 ab	23.0 ± 0.6 a	13.6 ± 0.4 a

CK indicates no biochar and no Cd in soil; B0, B4, B8, B16, and B32 indicate adding 0, 4, 8, 16, and 32% of biochar in soil, respectively; and different lower-case letters in the same column indicate significant differences ($P < 0.05$) among treatments.

Table 6. Effects of biochar on relative chlorophyll content of *S. miltiorrhiza* in Cd-polluted soil.

demonstrated that 3 mg/kg of Cd stressed the plants resulting in an increase of membrane permeability. Our data showed that adding biochar could reduce cell membrane conductivity, but no change trend is related to the biochar ratio.

3.2.3. MDA content

Figure 4 showed the effects of different proportions of biochar on MDA content of *N. nucifera* 'Taikong Lian 36' grown for 20 d in Cd-polluted water and *S. miltiorrhiza* grown for 30 d in Cd-polluted soil. The highest MDA content was seen in the treatment H8 and the lowest in H32 for lotus (**Figure 4A**). However, no significant differences in MDA content were found between treatments where both Cd and biochar were added, which indicated that the amount of Cd did not produce a great impact on the cell membrane lipid peroxidation. The other possible reason was that the physiological metabolic activity of *N. nucifera* 'Taikong Lian 36' was disturbed by Cd, leading to the accumulation of oxidation product MDA, while the MDA accumulation was inhibited by biochar, which alleviated the damage of the membrane system caused by lipid membrane peroxidation.

In Chinese sage, the highest MDA content was seen in CK (**Figure 4B**), which indicated that the amount of Cd did produce some impact on the lipid peroxidation of cell membranes. Less MDA content occurred in biochar treatments, and the lowest was seen in B16, although no differences were found among the treatments with both Cd and biochar. The possible reason for no difference in treatments might be that the physiological metabolic activity was disturbed by the MDA accumulation due to Cd addition, while the MDA accumulation was inhibited by biochar, which alleviated the damage of the membrane system caused by lipid membrane peroxidation.

3.2.4. SOD activity

When both *N. nucifera* 'Taikong Lian 36' and *S. miltiorrhiza* grew for 30 d in Cd-contaminated environment, the effects of biochar on SOD activity in fresh leaves reached a significant level

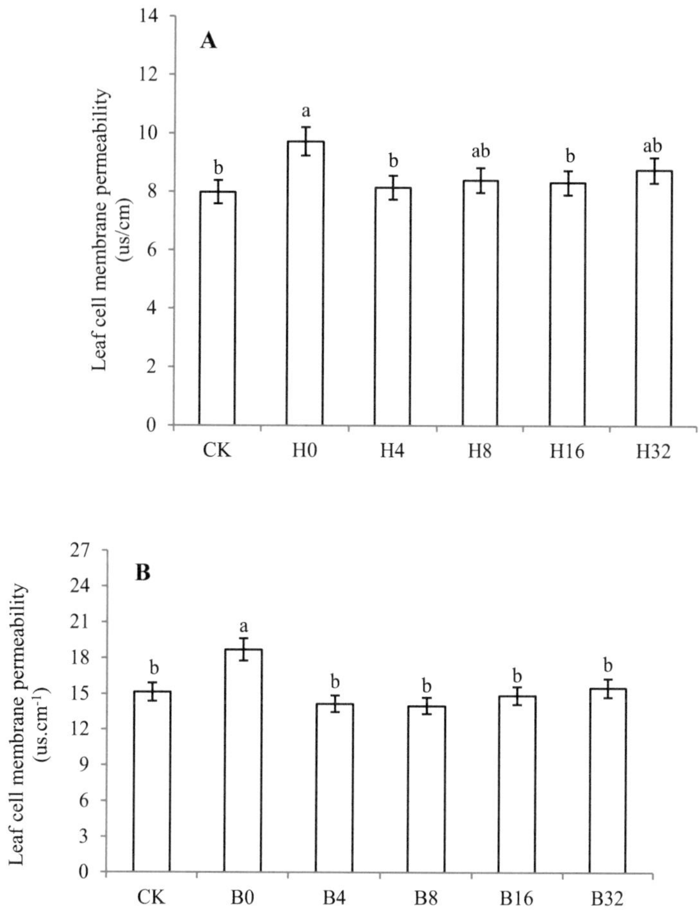

Figure 3. Effects of biochar on leaf cell membrane permeability of *N. nucifera* 'Taikong Lian 36' in Cd-polluted water (A) and *S. miltiorrhiza* in Cd-polluted soil (B). CK refers to none of biochar and Cd added to the soil; H0, H4, H8, H16, and H32 refer to adding 0, 4, 8, 16, and 32% of biochar in soil, respectively. Different lower-case letters indicate significant difference ($P < 0.05$) among treatments.

(**Figure 5**). The highest SOD activity appeared in treatment H0, which indicated that Cd significantly enhanced the SOD activity of plants. However, after adding biochar, SOD activity declined significantly, with the highest decrease by 53% in treatment H8 for lotus (**Figure 5A**) and by 29% in treatment B16 for Chinese sage (**Figure 5B**), comparing to the control H0, though a decrease trend was inconsistent with the proportion of biochar.

3.3. Cd content

The Cd accumulation in plants clearly increased after adding Cd but significantly decreased when biochar is added (**Figures 6** and **7**). For *N. nucifera* 'Taikong Lian 36', the Cd content in rhizomes, petioles, and leaves increased by 12.81, 17.35, and 7.56 folds, respectively, in treatment H0, compared with the control (**Figure 6**). With increasing biochar, Cd in plant tissue decreased correspondingly. In H32 treatment, the Cd content in rhizomes was 0.125 mg/kg,

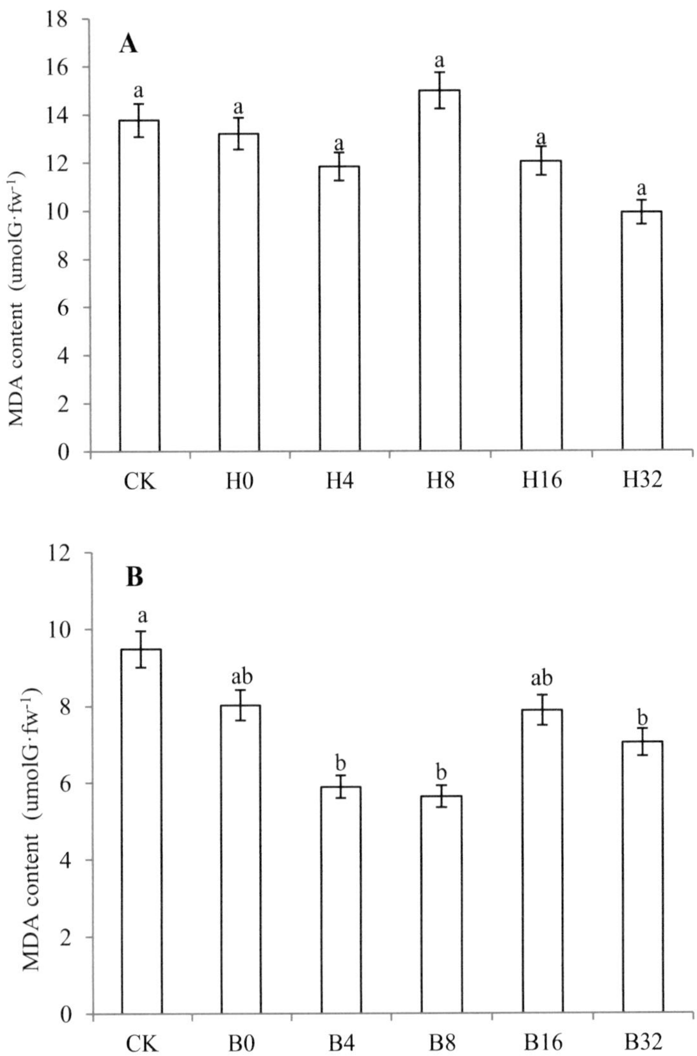

Figure 4. Effects of biochar on MDA content in leaf of *N. nucifera* 'Taikong Lian 36' in Cd-polluted water (A) and *S. miltiorrhiza* in Cd-polluted soil (B). CK refers to none of biochar and Cd added to the soil; H0, H4, H8, H16, and H32 refer to adding 0, 4, 8, 16, and 32% of biochar in soil, respectively. Different lower-case letters indicate significant difference ($P < 0.05$) among treatments.

only 68.51% of that in H0. It was more reduced in petioles from 0.694 to 0.129 mg/kg, a decrease of 81.41%. The lowest Cd content in leaves was 0.209 mg/kg in H32 and decreased by 54.66% of the highest in H0. However, the Cd content in all evaluated plant parts remained above the limit set by the European Commission regulations.

The relationship for the proportion of biochar (x) relative to Cd content (y) in rhizomes could be reflected by (5) $y = -0.0002\ x^3 + 0.008\ x^2 - 0.075\ x + 0.3999$ ($R^2 = 0.999$, $P = 0.351$). Based on this relationship, a higher level of biochar possibly further reduces Cd uptake and makes the

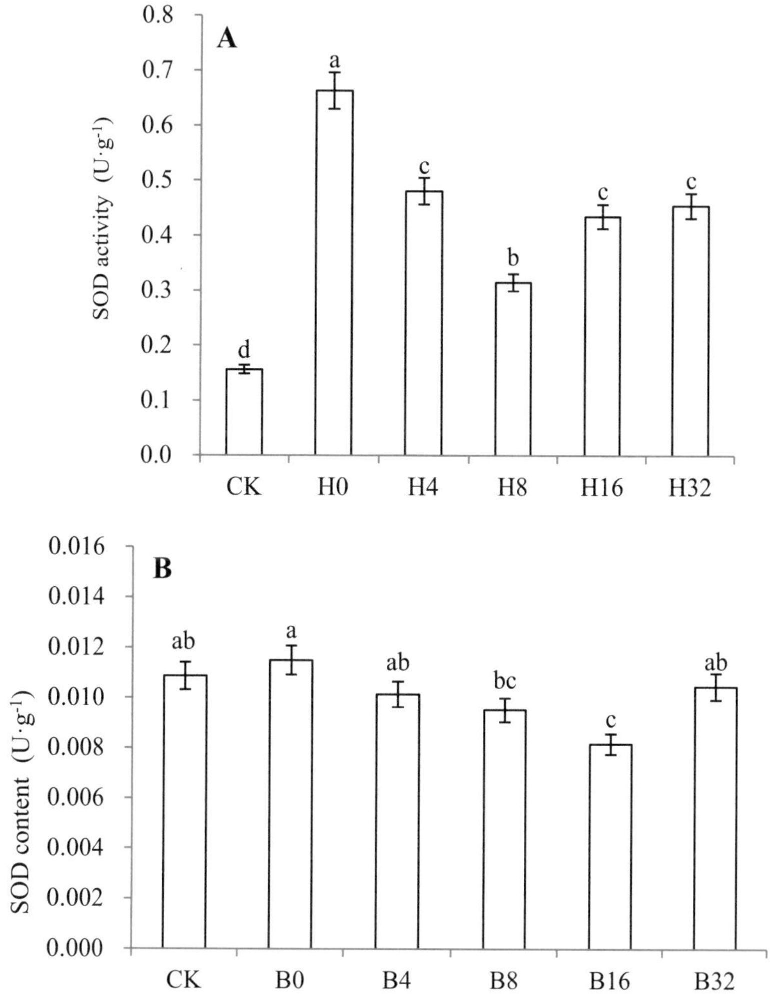

Figure 5. Effects of biochar on SOD activity in the leaf of *N. nucifera* 'Taikong Lian 36' in Cd-polluted water (A) and *S. miltiorrhiza* in Cd-polluted soil (B). CK refers to none of biochar and Cd added to the soil; H0, H4, H8, H16, and H32 refer to adding 0, 4, 8, 16, and 32% of biochar in soil, respectively. Different lower-case letters indicate significant difference ($P < 0.05$) among treatments.

rhizomes of lotus grown in Cd contaminated condition meet food safe standards. The relationship between the proportion of biochar (x) and Cd content (y) in petioles and leaves could be presented by (6) $y = -0.0003\,x^3 + 0.012\,x^2 - 0.142\,x + 0.674$ ($R^2 = 0.965$, $P = 0.238$) and (7) $y = -0.00004\,x^3 + 0.002\,x^2 - 0.047\,x + 0.489$ ($R^2 = 0.793$, $P = 0.559$), respectively.

In Chinese sage, Cd content in roots and leaves was 0.57 and 0.18 mg/kg, respectively, in treatment B0 and increased to 2.2 and 2.0 folds compared with the control (**Figure 7**). However, with increasing biochar, Cd content in plant tissues decreased correspondingly. In roots, Cd content was the lowest 0.25 mg/kg in B32, only 43.6% of that in B0. It was much lower in leaves

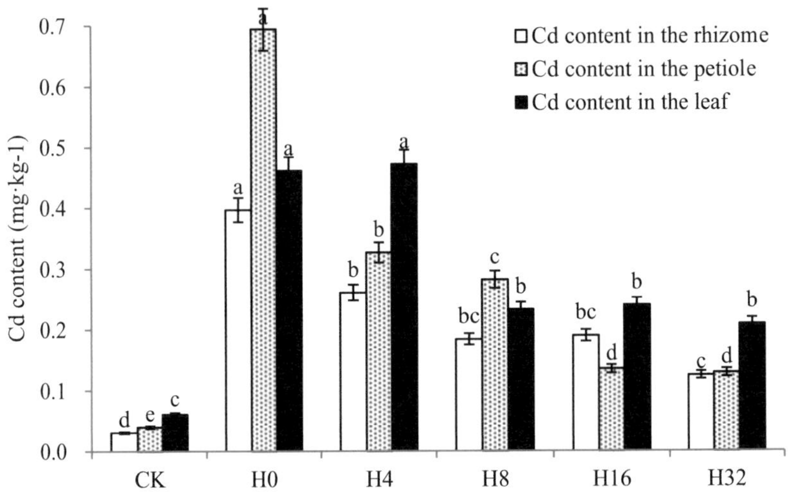

Figure 6. Effects of biochar on Cd content in the root, petiole, and leaf of *N. nucifera* 'Taikong Lian 36' in Cd-polluted water. CK refers to none of biochar and Cd added to the soil; H0, H4, H8, H16, and H32 refer to adding 0, 4, 8, 16, and 32% of biochar in soil, respectively. Different lower-case letters indicate significant difference ($P < 0.05$) among treatments.

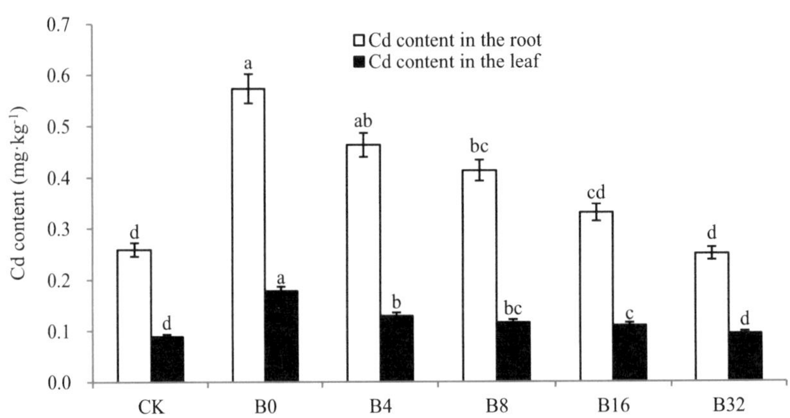

Figure 7. Effects of biochar on Cd content in the root and leaf of *S. miltiorrhiza* in Cd-polluted soil. CK indicates no biochar and no Cd in soil; B0, B4, B8, B16, and B32 indicate adding 0, 4, 8, 16, and 32% of biochar in soil, respectively. Different lower-case letters indicate significant differences ($P < 0.05$) among treatments.

in the same treatment, down to 0.09 mg/kg, a decrease of 52.8% of the highest in B0, and was below the upper limit value set by the *Chinese Pharmacopoeia* [22].

The regression equations for the proportion of biochar (x) related to Cd content (y) in roots and leaves, respectively, were

$$y = -0.00007x^3 + 0.003x^2 - 0.046x + 0.067 \; (R^2 = 0.997, \; P = 0.067) \quad (8)$$

$$y = -0.00007x^3 + 0.002x^2 - 0.023x + 0.177 \; (R^2 = 0.991, \; P = 0.119) \quad (9)$$

The increasing proportion of biochar in 3 mg/kg Cd-contaminated planting mix likely further prevents Cd uptake to make the roots (a major medicinal part) meet safe standards of Chinese medicine.

3.4. Cd BCF and S/R

The BCF shows the capacity of plants to adsorb heavy metals. In lotus case, the BCF was the lowest in the rhizomes of the control, only 7.6% of the highest value of H0 (**Figure 8**). However, after adding biochar, the BCF decreased significantly, with the lowest (0.042) in H32, about one-third of that in the control. The highest BCF (0.385) of the above-ground parts occurred in H0 and was 11 times greater than that in the control. Likewise, the above-ground BCF was reduced significantly, with the biggest decrease seen in H32, down to 70.65% of that in the control. The petiole's BCF was significantly less than that of the leaves.

The transfer coefficient from roots to shoots (S/R) showed that Cd was transported to different organs of plants. The S/R rose with the increasing biochar proportions and the highest value (3.06) was observed in H32. The increase of S/R was not significant, but it indicated that biochar could affect Cd transference and reduce its accumulation in rhizomes.

In Chinese sage, the highest BCF (0.19) of the roots occurred in B0, 2.2 times that of the control (**Figure 9**), indicating that roots were able to accumulate Cd. However, after adding biochar, the BCF of both root and leaf decreased with the rate of biochar. The lowest root BCF (0.08) in treatment B32 was only 43% of the highest observed in B0. Likewise, the lowest BCF of leaf

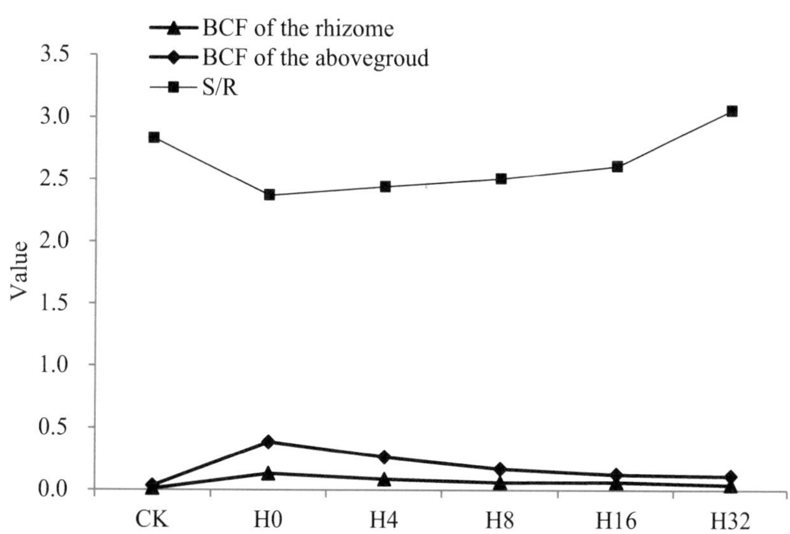

Figure 8. Effects of biochar on Cd bioconcentration factor (BCF) and biological transfer coefficient (S/R) of *N. nucifera* 'Taikong Lian 36' in Cd-polluted water. CK refers to none of biochar and Cd added to the soil; H0, H4, H8, H16, and H32 refer to adding 0, 4, 8, 16, and 32% of biochar in soil, respectively. Different lower-case letters indicate significant difference ($P < 0.05$) among treatments.

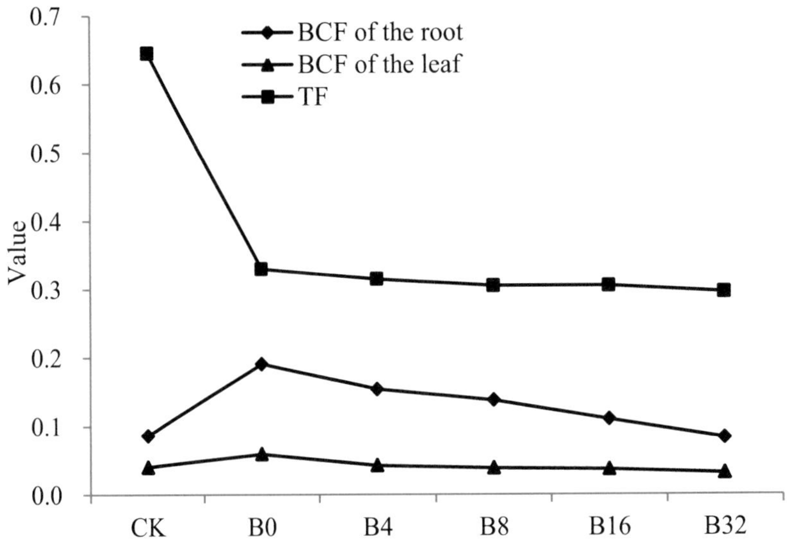

Figure 9. Effects of biochar on Cd bioconcentration factor (BCF) and biological transfer factor (TF) of *S. miltiorrhiza* in Cd-polluted soil. CK refers to no biochar and no Cd in soil; B0, B4, B8, B16, and B32 indicate adding 0, 4, 8, 16, and 32% of biochar in soil, respectively.

also occurred in B32, 77.5% of the highest in B0. It is possibly due to the capacity of biochar to inhibit absorption of Cd in plants.

The transfer factor (TF) from roots to leaves showed that Cd can be absorbed from soil by *S. miltiorrhiza* and transported among different plant organs. The highest TF was observed in the control (0.65). TF decreased with the addition of Cd. This could be explained by Cd potentially inhibiting growth and metabolism of *S. miltiorrhiza*, therefore slowing the absorbed Cd from the roots. With the increasing biochar proportion, TF decreased correspondingly. The lowest TF was 0.3 in B32, less than half of that in the control, indicating that biochar could affect the transfer of Cd from roots to leaves.

4. Discussion

4.1. Effects of biochar on plant growth

In this study, 3-mg/kg Cd added to the soil or soil-water mix did not elicit overt symptoms of heavy metal poisoning such as leaf chlorosis or stunting in lotus but resulted in treated plants being stunted and undersized in Chinese sage when compared to CK and biochar treatments. The lack of observed toxicity in *N. nucifera* may be due to its strong resistance of the polluted environments, such as in a Cd-polluted case study [23]. For both cases of our study, the leaf growth, total fresh weight, aboveground fresh weight, and fresh and dry weight of underground parts of plants increased with biochar ratio, and nearly all the highest values appeared in the treatment with the highest amount of biochar. Therefore, biochar can promote plant growth, which has also been reported in the previous studies [7, 24]. Because no fertilizer was applied after adding biochar, the biomass increase suggests that biochar might work as a fertilizer [25] or fertilizer enhancer [24], by ameliorating soil pH, CEC, base saturation, and other physicochemical properties, by affecting soil nutrient availability, such as the effectiveness of K, Ca, Na, Mg, N, P, and other elements [26], or by increasing the rhizosphere and soil nitrogen retention indirectly [27]. Yet some studies also showed that biochar had no beneficial effect on plant biomass, which is related to the type of soil and biochar, fertility levels, and other relevant experimental conditions [5, 28, 29]. Although Cd inhibited the underground growth of lotus, adding biochar offsets the stress of Cd and significantly promoted the rhizome enlargement and biomass increase. The 16–32% of biochar added in soil was more beneficial for growth of both lotus and Chinese sage. However, more biochar addition will increase production cost.

4.2. Effects of biochar on physiological status

Cell membrane permeability, relative chlorophyll content, MDA, and SOD served as markers to represent plant reactions under stress conditions. Our study showed that Cd added to water or soil produced a stress on cell membrane structure of plant, but no significant difference occurred among treatments with different proportions of biochar. Therefore, no optimal biochar level could be determined based on this point. Possibly, the beneficial effects of biochar on membrane permeability were also affected by other environmental factors, such as pH, EC, etc., making the membrane permeability change irregularly. The chlorophyll content in plant leaves increased with biochar amount in all treatments, which coincided with the previous report, in which low Cd concentration facilitated chlorophyll formation and high Cd concentration reduced chlorophyll content [23, 30]. This phenomenon also occurred in other heavy metal study cases [31]. In the early period of plant growth, promotion effect of biochar on chlorophyll was not evident. However, leaf chlorophyll content increased in the middle period, followed by a relatively big decrease in the latter. Therefore, biochar promoted plant growth and advanced plant senescence simultaneously. In lotus case, after adding Cd and biochar, the MDA contents showed no difference among treatments because low Cd concentrations did not increase cell membrane peroxidation. Therefore, a mitigating effect of biochar could not be determined. But, in the case of Chinese sage, the significant differences showed in MDA content between treatments with and without biochar, which suggested that biochar alleviated cell membrane peroxidation caused by Cd. In both cases, SOD activity significantly increased by adding Cd, indicating that Cd stimulated the physiological defense mechanisms and physiological resistance activities and improved activity of antioxidant and its enzyme [30, 32]. However, the addition of biochar significantly decreased SOD activity, which suggested that biochar can ameliorate Cd hazards.

4.3. Effects of biochar on Cd content in plants

Cd accumulation was observed in the organs and tissues of *N. nucifera* 'Taikong Lian 36' and *S. miltiorrhiza* in artificially polluted environment, and the accumulative concentrations increased more than 10 folds in some treatments when compared with the control. The uptake of Cd was significantly reduced by adding biochar in all treatments, but the maximum effect was observed in H32. The Cd content in roots of *S. miltiorrhiza* could be reduced to meet the maximum allowed value set by the *Chinese Pharmacopoeia* [22]. In both cases, Cd content in plants could possibly be further decreased with more biochar added, extrapolating from the regression equation of biochar and Cd content. The role of biochar on reducing the heavy metal accumulation in plants has been demonstrated in the previous reports [33, 34]. It might be related to huge specific surface area and strong adsorption functions of biochar to form insoluble chelates or cause Cd precipitation. Biochar reduced the bioavailability of Cd in soil or water, by affecting pH, CEC (cation exchange capacity), water-holding capacity, and other physicochemical properties of the soil. Although the Cd-contaminated lotus rhizomes in our study did not meet the edible standard, the regression equation of biochar and Cd content indicated that it was possible to further reduce the Cd content to the safe range if more biochar would be applied or the environment is less polluted. Combined with Cd contents in different organs of *N. nucifera* 'Taikong Lian 36', all the lowest BCFs in various organs occurred in the controls, while the highest in treatments H0. In Chinese sage, BCFs of the root and leaf decreased with the addition of biochar indicate that besides its strong ability of adsorbing Cd, biochar also reduced the bioavailability of heavy metals or immobilized heavy metals by changing the physicochemical property of soil [35]. The highest S/R values were observed in the treatment with 32% biochar (H32), and the lowest showed in the treatment H0. Based on the values of Cd content, Cd BCF, and S/R, 32% of biochar was the optimum amount for applying to demonstrate biochar's potential in plant growth promotion, ventilation and water retention, fertilizer efficiency, and pollution reduction.

5. Conclusion

In this study, the effects of biochar were evaluated on edible and medicinal plants. The major goals were to understand growth-promoting effects of biochar and investigate how biochar alleviates the toxicity and accumulation of Cd in artificially Cd-amended pot culture. The results showed that the contents of Cd in rhizomes, petioles, and leaves of *N. nucifera* 'Taikong Lian 36' and in roots, leaves, and stems of *S. miltiorrhiza* were significantly reduced and the transference of Cd in plants was inhibited by adding biochar. Although the impacts of biochar on plant physiological indicators were not very clear, biochar not only promoted plant growth but also alleviated toxic effects of Cd in Cd-contaminated environment. Therefore, as a valuable soil amendment, biochar has a huge potential in green agriculture production and remediation of heavy metal-polluted water or soil due to its positive effects of improving plant growth and reducing heavy metal uptake. Further studies are needed to evaluate effects of biochar on safe production of lotus in the field and other important vegetables, fruit trees, and Chinese medicinal plants and to establish the optimum of physiologically and economically feasible amount of biochar to add for environmental remediation.

Acknowledgements

This study was funded by Shanghai Administration Bureau of Landscape and City Appearance.

Author details

Daike Tian[1,2]*, Amei Liu[1,2] and Yanci Xiang[3]

*Address all correspondence to: dktian@sibs.acn.cn

1 Shanghai Chenshan Plant Science Research Center, Chinese Academy of Sciences, Shanghai, China

2 Shanghai Key Laboratory for Plant Functional Genomics and Resources, Shanghai Chenshan Botanical Garden, Shanghai, China

3 School of Life Science, Hunan University of Science and Technology, Xiangtan, Hunan, China

References

[1] Beesley L, Moreno-Jiménez E, Gomez-Eyles JL. Effects of biochar and green waste compost amendments on mobility, bioavailability and toxicity of inorganic and organic contaminants in a multi-element polluted soil. Environmental Pollution. 2010; **158**(6):2282-2287

[2] Lohmann R, Macfarlane JK, Gschwend PM. Importance of black carbon to sorption of native PAHs, PCBs, and PCDDs in Boston and New York harbor sediments. Environmental Science and Technology. 2005;**39**:141-148

[3] Lu H, Li Z, Fu S, Méndez A, Gascó G, Paz-Ferreiro J. Effect of biochar in cadmium availability and soil biological activity in an anthrosol following acid rain deposition and aging. Water Air Soil Pollution. 2015;**226**:164-174

[4] Mohan D, Pittman Jr. CU, Bricka M, Smith F, Yancey B, Mohammad J, Steele PH, Alexandre-Franco MF, Gómez-Serrano V, Gong H. Sorption of arsenic, cadmium, and lead by chars produced from fast pyrolysis of wood and bark during bio-oil production. Journal of Colloid and Interface Science. 2007;**310**:57-73

[5] Alburquerque JA, Salazar P, Barrón V, Torrent J, del Campillo MC, Gallardo A, Villar R. Enhanced wheat yield by biochar addition under different mineral fertilization levels. Journal of Plant Nutrition and Soil Science. 2013;**177**:16-25

[6] Ventura F, Salvatorelli F, Piana S, Pieri L, Pisa PR. The effects of biochar on the physical properties of bare soil. Earth and Environmental Science Transactions of the Royal Society of Edinburgh. 2012;**103**(01):5-11

[7] Huang M, Yang L, Qin HD, Jiang LG, Zou YB. Fertilizer nitrogen uptake by rice increased by biochar application. Biology and Fertility of Soils. 2014;**50**(6):997-1000

[8] Steinbeiss S, Gleixner G, Antonietti M. Effect of biochar amendment on soil carbon balance and soil microbial activity. Soil Biology & Biochemistry. 2009;**41**(6):1301-1310

[9] Sika MP, Hardie AG. Effect of pine wood biochar on ammonium nitrate leaching and availability in a South African sandy soil. European Journal of Soil Science. 2014;**65**(1):113-119

[10] Reverchon F, Flicker RC, Yong H, Yan GJ, Xu ZH, Chen CR. Changes in $\delta^{15}N$ in a soil-plant system under different biochar feedstocks and application rates. Biology and Fertility of Soils. 2014;**50**(2):275-283

[11] Uchimiya M, Lima IM, Thomas Klasson K, Chang S, Wartelle LH, Rodgers JE. Immobilization of heavy metal ions (Cu^{II}, Cd^{II}, Ni^{II}, and Pb^{II}) by broiler litter-derived biochars in water and soil. Journal of Agricultural Food Chemistry. 2010;**58**(9):5538-5544

[12] Liu HQ. The sadness of cadmium-rice. The Earth. 2013;**06**:64-65 (in Chinese)

[13] Qiu LL. Face the crisis of cadmium-rice. China Food. 2011;**06**:74-75 (in Chinese)

[14] Zhang XY, Chen LQ, Wang QC. New Lotus Flower Cultivars in China—I. 1st ed. Beijing, China: China Forestry Publishing House; 2011. p. 279

[15] Li MX. Effects of cadmium on the growth and development of lotus (*Nelumbo nucifera* Gaertn) [thesis]. Wuhan, China: College of Horticulture & Forestry Sciences, Huazhong Agricultural University; 2005

[16] Xiong CH, Xu XG, Lu YE, Ou YB, Zhang YY, Ye ZB, Li HX. Effects of heavy metals Pb and Cd on the physiological response and accumulation in lotus root. Acta Horticulturae Sinica. 2012;**39**(12):2385-2394

[17] Han JY, Fan JY, Horie Y, Miura S, Cui DH, Ishii H, Hibi T, Tsuneki H, Kimura I. Ameliorating effects of compounds derived from *Salvia miltiorrhiza* root extract on microcirculatory disturbance and target organ injury by ischemia and reperfusion. Pharmacology & Therapeutics. 2008;**117**:280-295

[18] Pan XJ, Niu GG, Liu HZ. Microwave-assisted extraction of tanshinones from *Salvia miltiorrhiza* bunge with analysis by high-performance liquid chromatography. Journal of Chromatography A. 2001;**922**:371-375

[19] Xiong QE, editor. Plant Physiology Experiments Tutorial. 1st ed. Chengdu, China: Sichuan Science and Technology Press; 2003. pp. 124-127

[20] Chinese Ministry of Health. Determination of Cadmium in Foods, GB/T 5009.15-2003. Beijing, China: Standardization Administration of China; 2003

[21] Zayed A, Gowthaman S, Terry N. Phytoaccumulation of trace elements by wetlands plants: I. Duckweed. Journal Environmental Quality. 1998;**27**:715-721

[22] Chinese Pharmacopoeia Commission. Chinese Pharmacopoeia. Beijing, China: China Medical Science Press; 2010

[23] Pei KK. Effect of Cd and Zn stress on seed germination, physiology and biochemistry of *Nelumbo nucifera* [thesis]. Zhengzhou, China: College of Forestry, Henan Agricultural University; 2010 (in Chinese with English abstract)

[24] Utomo Widowati WH, Guritno B, Soehoo LA. The effect of biochar on the growth and N fertilizer requirement of maize (*Zea mays* L.) in green house experiment. Journal of Agricultural Science. 2012;**4**(5):255-262

[25] Lehmann J. Bio-energy in the black. Frontiers in Ecology and the Environment. 2007;**5**(7): 381-387

[26] Oguntunde PG, Fosu M, Ajayi AE, Gesen N. Effects of charcoal production on maize yield, chemical properties and texture of soil. Biology and Fertility of Soils. 2004; **39**(4):295-299

[27] Prendergast-Miller MT, Duvall M, Sohil SP. Biochar-root interactions are mediated by biochar nutrient content and impacts on soil nutrient availability. European Journal of Soil Science. 2014;**65**(1):173-185

[28] Devereux RC, Sturrock CJ, Mooney SJ. The effects of biochar on soil physical properties and winter wheat growth. Earth and Environmental Science Transactions of the Royal Society of Edinburgh. 2012;**103**(01):13-18

[29] Eyles A, Bound SA, Oliver G, Corkrey R, Hardie M, Green S, Close DC. Impact of biochar amendment on the growth, physiology and fruit of a young commercial apple orchard. Trees. 2015;**29**(6):1817-182

[30] Yang SC, Wu Q. Effect of Cd on growth and physiological characteristics of *Aegiceras conrniculatum* seedlings. Marine Environmental Science. 2003;**22**(1):38-42 (in Chinese with English abstract)

[31] Jonker MT, Koelmans AA. Sorption of polycyclic aromatic hydrocarbons and polychlorinated biphenyls to soot and soot-like materials in the aqueous environment: Mechanistic considerations. Environmental Science and Technology. 2002;**36**(17):3725-3734

[32] Li ZZ, Wu J, Tang Y, Yang G. Effect of Pb, Zn and their interactions on the chlorophyll content and antioxidant enzyme systems of *Houttuynia cordata* Thunb. Acta Ecologica Sinica. 2007;**27**(12):5441-5446

[33] Fellet G, Marchiol L, Delle Vedove G, Peressotti A. Application of biochar on mine tailings: Effects and perspectives for land reclamation. Chemosphere. 2011;**83**(9):1262-1267

[34] Puga AP, Abreu CA, Melo LCA, Paz-Ferreiro J, Beesley L. Cadmium, lead, and zinc mobility and plant uptake in a mine soil amended with sugarcane straw biochar. Environmental Science and Pollution Research. 2015;**22**(22):17606-14

[35] Rees F, Simonnot MO, Morel JL. Short-term effects of biochar on soil heavy metal mobility are controlled by intra-particle diffusion and soil pH increase. European Journal of Soil Science. 2014;**65**(1):149-161

Chapter 5

Biochar Adsorption Treatment for Typical Pollutants Removal in Livestock Wastewater: A Review

Yaxin Deng, Tao Zhang and Qiming Wang

Additional information is available at the end of the chapter

http://dx.doi.org/10.5772/intechopen.68253

Abstract

Biochar, as an high efficiency, environmental friendly, and low-cost adsorbent, is usually used as soil conditioner, bio-fuel, and carbon sequestration regent. Recently, biochar has attracted much attention in wastewater treatment field. There are plenty of studies about application of biochar to adsorb pollutants in wastewater, because of its low-cost preparation, high surface area, large pore volume, plentiful functional groups, and environmental stability. Furthermore, it can be reused due to their high treatment efficiency and resource recovery potential. As biochar can be used for adsorption of typical pollutants in livestock wastewater, it becomes a promising method to treat livestock wastewater. The preparation methods, including pyrolysis, hydrothermal carbonization, and gasification, were introduced. The applications of biochar to adsorb typical pollutants, such as organic pollutants, heavy metals, and nutrients, in livestock wastewater were present. The organic structures, surface functional groups, surface electricity, and mineral component of biochar were investigated to explain the adsorption mechanism of organic pollutants, heavy metals, and nutrients in wastewater. Finally, outlooks were made for the better use of biochar in future. The relationship of preparation parameters, structures, and adsorption performance of biochar should be discussed. The quantitative analysis for the adsorption of organic structures, surface functional groups, surface electricity, and mineral component should be performed. The disposal of post-sorption biochar should be investigated.

Keywords: biochar, adsorption, livestock wastewater, review

1. Introduction

Biochar is a carbon-rich solid product formed by pyrolysis of bio-organic materials at middle to low temperature (<700°C) under anoxic conditions [1–3]. The raw materials of biochar are mainly biomass waste (straw, feces, or sludge), which not only benefits waste resource utilization but also effectively alleviate environmental deterioration [4]. The methods to make biochar, including pyrolysis, hydrothermal carbonization, gasification, and so on, different bio-organic materials, pyrolysis temperature, modification methods, and other factors will influence the performance of biochar. The most studies mainly focus on these areas such as soil performance regulation [5, 6], global climate change [7], and renewable biofuels [8]. In recent years, biochar in the field of wastewater treatments has become a hot research [9–11].

Large-scale farming is an important direction of agricultural economic development and an important way for farmers to increase income [12]. However, large-scale farming will produce large amounts of wastewater; especially, the excessive emissions of organic pollutants, heavy metals, nitrogen, and phosphorus will cause serious contamination of the water environment. The first national pollutant source general survey technical report shows: agricultural non-point pollutant source is the main reason for water pollution in China [13]. At present, the harmless treatment of typical pollutants in livestock wastewater has become the focus of agricultural development and environmental protection.

At present, biochar adsorption technology is gradually applied in the field of wastewater treatment. It has been shown that biochar has good adsorption effect on typical pollutants of livestock wastewater such as organic pollutants [10], heavy metals [14], nitrogen, and phosphorus [11]. After magnetization, biochar with good magnetic is easily separated from liquid [15], which is more suitable for livestock wastewater, compared to commercial activated carbon products. At the same time, because biochar has good adsorption capacity for nitrogen and phosphorus, it can be used as a slow-release fertilizer and has the characteristics of agricultural environment-friendly [16].

In this paper, the preparation methods and structures of biochar, the efficiency, and the mechanism of adsorption of typical pollutants in livestock wastewater were reviewed. The mechanism of biochar adsorption of typical pollutants was discussed in organic carbon structure, surface functional groups, surface electrical properties, and mineral components. The preparation condition, raw material characteristics, biochar structure, and adsorption properties were discussed, including their mutual relationships. At the same time, the quantitative mechanism of pollutant adsorption and the application of biochar were also discussed.

2. Biochar preparation methods and characteristics

2.1. Preparation methods

Biochar preparation methods are mainly pyrolysis, hydrothermal carbonization, gasification, and other methods (**Table 1**). Pyrolysis is an decomposition reaction under high temperature and anoxic conditions. Based on pyrolysis time, temperature, and heating rate, it can be divided into slow pyrolysis, rapid pyrolysis, and "flash" pyrolysis (heating rate up to 1000°C/s) [17]. Studies have shown that lower pyrolysis temperatures and slower heating rates contribute to the formation of solid products, as shown in **Table 1**, in the slow thermal cracking, the content of solid product up to 35%, and therefore, slow pyrolysis is considered as a main preparation of biochar [18]. Hydrothermal carbonization (HTC) refers to the reaction of

Preparation methods	Temperature (°C)	Heating rate	Residence time	Yield (%)		
				Solid	Liquid	Gas
Pyrolysis						
Slow pyrolysis	<700	Slow	h	35	30	35
Fast pyrolysis	<1000	Fast	s	10	70	20
Flash pyrolysis	775–1025	Faster	s	10–15	70–80	5–20
Hydro-carbonization	<350	Slow	min-h	50–80	–	–
Gasification	700–1500	Faster	s-min	10	5	85

s, second; min, minute; h, hour.

Table 1. Comparison of different techniques to make biochar [19, 24–26].

biomass in an underwater stagnant system for 5 min to 16 h at a pressure of 2–6 MPa and a relatively low temperature (<350°C) [18–20]. Because the hydrothermal carbonization process uses water as the reaction medium under high pressure and heating conditions, it is not easy to produce harmful substances. Therefore, the biochar prepared by this method is more suitable for the adsorption of water pollutants [21–23]. But this method is limited by the preparation conditions, the need for high pressure and high temperature of the expensive reactor. Because of its high preparation cost [21], the practical application is difficult to popularize. Other methods, such as drying, gasification, rapid pyrolysis, and "flash" pyrolysis, are mainly used to produce bio-oil or gaseous materials [19] because of the relatively small yield of solid products obtained, such as gas-product content of gasification about 90%, in which the higher the temperature the higher the content of gaseous products.

2.2. Characteristics

Biochar, with high carbon content and void structures, has abundant aromaticity oxygen-containing functional groups. The physic-chemical properties of biochar vary with the types of raw material, the particle size of the feedstock, the means of pyrolysis, the temperature (including the rate of temperature rise), the time of pyrolysis, and the modification conditions [19, 27, 28]. Although the structure of biochar is affected by many factors, in general, biochar has abundant surface functional groups (hydroxyl, carboxyl, carbonyl, and methyl) [29], the developed pore structure, the high specific surface area, and the stable molecular structure [30], with good adsorption performance, which is favor to adsorb pollutants in livestock wastewater.

3. Typical pollutants treatment

Livestock wastewater contains large amounts of organic matter, heavy metals, nitrogen, phosphorus, and other typical pollutants, causing serious harm to the environment. Biochar has a strong pollutants adsorption in liquid phase. Tan et al. [19] summarized the applications of biochar in adsorption of water pollutants, 39% of which for the adsorption of organic pollutants, 46% for heavy metal adsorption treatment, 13% for the adsorption of nitrogen and phosphorus, other studies accounted for only 2%.

3.1. Organic pollutants

Biochar has significant adsorption effect for organic pollutants such as antibiotics, phenols, herbicides, etc. [31]. Due to the similarity between the type of pollutants adsorbed and the types of organic contaminants in livestock wastewater, it has gained attention in agricultural resources and the environment.

Biocarbon can adsorb antibiotic substances in water phase (fluoroquinolone, sulfamethoxazole, etc.) and its adsorption mainly through π-π electron donor/receptor, hydrogen bonding, and cationic bridge. Yao et al. [32] prepared the biochar by pyrolysis of sludge for 1 h at 500°C, whose maximum adsorption capacity for fluoroquinolone (an antibiotic, clinical for the treatment of infection of urinary tract, intestinal, respiratory and skin soft tissue, abdominal cavity, and joint) was 19.80 ± 0.40 mg/g. And it was found that the content of volatile matter in the source sludge was positively correlated with the adsorption amount of fluoroquinolones by biochar. Zheng et al. [33] used biochar from donax to adsorb sulfamethoxazole (an antibiotic, curing acute and chronic urinary tract infection caused by *Escherichia coli* and Proteus), and inorganic components in the raw material enhanced the adsorption capacity of sulfamethoxazole in low-temperature pyrolysis biochar, and weakened adsorption capacity of sulfamethoxazole in the high-temperature pyrolysis biochar.

Biochar has significant adsorption effects on the high-chroma organic pollutants [34, 35], phenols, herbicides, etc. in the aqueous phase, and its adsorption mechanism involves a variety of physical and chemical effects, mainly depending on the polarity of organic pollutants and biochar, aromatic or matching property of special functional groups. Its physical adsorption mainly depends on the function of the electrostatic force and intermolecular gravitation between biochar and organic pollutants. And chemical adsorption is mainly through the chemical interactions between biochar and organic pollutants establishing the hydrogen bonds, π bonds, and coordination bonds. Xu et al. [36] used cole, peanut, rapeseed straw as raw materials to prepare biochar pyrolyzed at 350°C, whose methyl violet adsorption capacity of 123.5–195.4 mg/g. And the adsorption of methyl violet on the biochar from rapeseed straw was the highest at room temperature. Zeta potential and FTIR analysis showed that there was electrostatic attraction between methyl violet and biochar. The adsorption of methyl violet on —COO— and hydrophilic was dominant. Sun et al. [37] respectively obtained biochar from anaerobic digestion residue, palm bark and tree, through pyrolysis at 400°C for 30 min, respectively. Under the condition of 40°C, pH 7, 4 mg/L methyl blue, the removal efficiency was 99.5, 99.3 and 86.1%, respectively. The results showed that the pyrolysis temperature had a great effect on the removal efficiency of methyl blue. Lang et al. [38] prepared biochar from wheat straw and peanut shells at 300, 400, and 600°C, respectively. The results showed that the maximum adsorption capacity of wheat straw biochar and peanut shell biochar was up to 20.61 and 58.82 mg/g, respectively. Zheng et al. [39] prepared biochar from mixed wood waste pyrolyzed at http://dx.doi.org/10.5772/intechopen.68253

450°C for 1 h. The maximum adsorption capacities for atrazine (herbicide) and simazine (herbicide) were 1158 and 1066 mg/g, and the adsorption performance was better under acidic conditions. **Table 2** summarizes the studies about the biochar adsorption of organic pollutants.

3.2. Heavy metal pollutants

Heavy metals are toxic and cannot be biodegradable. Even if the concentration is low, it will pose threats to human health. Excessive emissions of heavy metals, such as Cu, Zn, Pb, and Cd, are found in livestock wastewater and can cause serious environmental pollution [44].

The adsorption of heavy metal ions on biochar mainly depends on the ion exchange on the surface of biochar, the chemical cross-linking between heavy metal ions and its surface functional groups and the surface deposition between the ashes. Inyang et al. [45] found that the removal efficiency of Ni(II),Cu (II), Pb (II), and Cd (II) in biochar at 22°C, which pyrolyzed from anaerobic digestion of sugar beet root at 600°C for 2 h, up to 97%. However, the adsorption capacity of the four kinds of ions was decreased, where the adsorption of Cd (II) was strongest, while the adsorption selectivity to Cu (II) was the weakest. Zhang et al. [46] found that the maximum adsorption of Cr (VI) on biochar of wheat straw decreased with the increase of pyrolysis temperature, and the maximum adsorption capacity of Cr (VI) was obtained when the pyrolysis temperature was 200°C at 35.78 mg/g. **Table 3** summarizes the studies about the biochar adsorption of heavy metal pollutants, which shows that biochar adsorption of heavy metals mainly depends on the raw materials, preparation conditions, adsorption temperature, and other conditions.

3.3. Nitrogen and phosphorus pollutants

Livestock wastewater contains a lot of nutrients especially nitrogen and phosphorus. The use of biochar for adsorption and fixing not only helps alleviate eutrophication but also can be recycled, re-applied to the soil, enhancing soil fertility, and recycling nutrient resources, which has been the research focus of the current resource recycling and reuse. Zhang et al. [52] found that the maximum adsorption capacity of ammonia nitrogen of biochar from corn cob pyrolyzed at 600°C for 2 h was up to 9.67 mg/g. Ma et al. [53] found that the maximum adsorption capacity of ammonia nitrogen of cow dung biochar was up to 25.84 mg/g at 25°C. Cheng et al. [54] prepared biochar by pyrolyzing municipal sludge anaerobic fermentation residue. The

Typical pollutants	Raw material	Pyrolysis temperature (°C)	Pyrolysis time (h)	Adsorption temperature (°C)	Maximum adsorption	References
Dimethyl sulfide	Chicken manure	450	1	22–24	–	[40]
Atrazine	Sludge	400	2	25 ± 2	27.03 μmol/g	[41]
Quinolone antibiotics	Bamboo	500	–	25 ± 2	45.88 mg/g	[42]
Pentachlorophenol	Paper-making sludge	700	–	15–40	47–50 mg/g	[43]

Table 2. Adsorption of organic pollutants in wastewater by biochar.

Typical pollutants	Raw material	Pyrolysis temperature (°C)	Pyrolysis time (h)	Adsorption temperature (°C)	Maximum adsorption	References
Hg(II)	Bagasse/pecan skin	450	2	25	13 mg/g	[13]
Mo	Microalgae (containing iron)	300, 450, 750	1	15	78.5 mg/g	[47]
As				15	62.5 mg/g	
Pb	Pruning branches of fruit trees	600	4	25	17.7–19.2 mg/g	[48]
As(II)	Sludge	400	2	–	3.08–6.04 m/g	[49]
Cd(II)	Water hyacinth	450	1	25 ± 1	70.3 mg/g	[50]
Cr(VI)	Wheat straw	350–650	–	25	23.6 mg/g	[51]

Table 3. Adsorption of heavy metal pollutants in wastewater by biochar.

results showed that sludge fermentation was conducive to the development of structure of biochar pores and improved the adsorption of nitrogen and phosphorus which fitting Langmuir model. Fang et al. [55] pyrolyzed corncob to prepare biochar and modified it with calcium and magnesium. The results showed that the modified biochar had a strong adsorption capacity to phosphate, and the maximum adsorption capacity was 319.63 mg/g, indicating that the use of calcium and magnesium cations to modify the biochar of corncob can enhance the anion exchange capacity of biochar, thereby enhancing its phosphate adsorption efficiency. Based on the results, it was demonstrated that the phosphorus absorbed in biochar is effective, which can be applied as fertilizer.

4. Biochar adsorption mechanism

In order to improve the adsorption efficiency of biochar on pollutants, especially for the typical pollutants in livestock wastewater, moreover, it is variable for the different type and properties of pollutants; therefore, it is very important to analyze the adsorption mechanism of biochar on the pollutants. In this paper, the adsorption mechanism of biochar is discussed from four aspects, including organic structure, surface functional group, surface electrical property and mineral composition.

4.1. Organic structure

The organic structure of the biochar is composed of two layers: stacked layers of graphene and aromatic structures which are interspersed with the graphene layer [56], armoring the biochar with the characteristics of large specific surface areas and rich pore structures. The large specific surface areas enhance the physical adsorption capacity of biochar, and the rich pore structures help to adsorb the organic matter with the same molecular weight [19]. Wang et al. [42] showed that biochar made by bamboo had a lot of pore structures, mesoporous structure accounted for nearly 90% of total pore structure, and quinolone antibiotics adsorption mainly may occurred in the mesoporous structures.

4.2. Surface functional groups

The functional groups (hydroxyl, carboxyl, etc.) on the surface of biochar fix metals by electrostatic attraction, complexation, and surface precipitation. Zhang et al. [57] showed that when the pyrolysis temperature increased to 500°C or higher, =OH and =CH on the surface of sludge biochar would be destroyed, although the more specific surface areas prepared on high-temperature conditions, the adsorption capacity of Pb (II) was weaker, indicating that surface functional groups =OH and =CH play major roles in the adsorption of heavy metals. Nguyen and Lee [40] found that the surface of chicken manure biochar modified by HNO_3/ NH_3 could form new amino functional groups, which could improve the adsorption performance to dimethyl sulfide.

4.3. Surface electrical properties

The electrostatic attraction ability on the surface of biochar plays a very important role in the adsorption of pollutants. In general, the surface electricity of biochar is negative, so it has a good adsorption performance for positive ions such as ammonia, heavy metals. If the biochar is modified so that the surface electricity is positively charged, anions such as phosphate can be adsorbed. Zhang et al. [47] found the biochar had a good adsorption effect on ammonia nitrogen, while the adsorption of phosphate was very weak. Fang et al. [11] modified corncob biochar with magnesium salt, so that the surface electricity of biochar was positive, which enhanced its phosphate adsorption efficiency.

4.4. Mineral ingredients

Mineral components such as CO_3^{2-}, PO_3^{2-}, etc. in biochar can increase their adsorption properties. Inyang et al. [45] found that the main role of Pb (II) adsorption by digested cow dung biochar was surface precipitation. Pb (II) reacted with CO_3^{2-}, HCO_3^-, $H_2PO_4^-$ ions on the surface of biochar forming $PbCO_3$, $Pb(CO_3)_2(OH)_2$ and $Pb_5(PO_4)_3X$\{S\} (where X may be F-, Cl-, Br-, or OH-) precipitation.

Of course, it is the variable synergies effects that adsorb specific pollutants. In general, the adsorption of organic pollutants by biochar is mainly through the combination of pore-immobilization and electrostatic attraction of organic functional groups. The adsorption of heavy metals mainly through electrostatic attraction, ion exchange, and complexation reaction of surface functional groups, as well as the precipitation of mineral components, the adsorption of nitrogen and phosphorus is mainly through the combination electrostatic attraction with precipitation of the mineral composition.

5. Conclusion and prospect

The above researches show that biochar, as a new type of adsorbent with high efficiency and environmental protection, has broad prospects in the field of adsorption treatment for typical pollutants in livestock wastewater. However, there are few studies on the relationship between raw materials and process parameters—biochar pore structures-adsorption properties, quantitative analysis of biochar adsorption mechanism of pollutants, and disposal of biochar after adsorption. The above problems can be further studied from the following three aspects.

1. The contribution of different raw materials and preparation parameters to the formation of biochar pore structures is not quantitatively described, and it plays strong guiding role for the adsorption of biochar, especially for organic pollutants, so it is necessary to study the relationship between biochar adsorption performance, biochar pore structures, and preparation parameters.
2. The adsorption mechanism of biochar on pollutants is the result of synergism of many kinds of adsorption processes, such as electrostatic attraction, complexation reaction, and precipitation reactions. Quantifying the effect of various adsorption processes on the total amount of pollutants can help to clarify the adsorption mechanism of biochar. Although a small number of quantitative models for organic adsorption mechanism, but for heavy metals, nitrogen and phosphorus adsorption qualitative and quantitative models need to explore in depth.
3. There are few researches on biochar recycling or regeneration after adsorption of pollutants. Although biochar can be used as bio-fertilizers or soil conditioners after adsorption of nitrogen and phosphorus, the biochar after adsorption of organic pollutants or heavy metal could lead to secondary pollution if it is not properly treated. Therefore, it is necessary to carry out further researches on the regeneration of the biochar to avoid environmental risks.

Acknowledgements

The work was supported by a grant from the National Natural Science Foundation of China (no. 31401944), the National Key Technology Research and Development Program of China (no. 2016YFD0501404-6), the China Agricultural University Education Foundation "Da Bei Nong Group Education Foundation" (no. 1031-2415005), the Specialized Research Fund for the Doctoral Program of Higher Education of China (no. 20120008120013), and the Beijing Municipal Natural Science Foundation (no. 6144026).

Author details

Yaxin Deng, Tao Zhang* and Qiming Wang

*Address all correspondence to: taozhang@cau.edu.cn

Key Laboratory of Farmland Pollution Prevention Control and Remediation of Beijing, Key Laboratory of Plant-Soil Interactions of Ministry of Education, College of Resources and Environmental Sciences, China Agricultural University, Beijing, China

References

[1] Lehmann J, Gaunt J, Rondon M. Bio-char sequestration in terrestrial ecosystems—A review. Mitigation and Adaptation Strategies for Global Change. 2006;**11**(2):395–419

[2] Biochar for environmental management: science and technology. Routledge, 2015.

[3] International Biochar Initiative. Terms and definition, International Biochar Initiative. http://www.biochar-international.org/definitions

[4] Chen B, Chen Z. Sorption of naphthalene and 1-naphthol by biochars of orange peels with different pyrolytic temperatures. Chemosphere. 2009;**76**(1):127–133

[5] Lehmann J, Rillig MC, Thies J, et al. Biochar effects on soil biota–a review. Soil Biology and Biochemistry. 2011;**43**(9):1812–1836

[6] Chan KY, Van Zwieten L, Meszaros I, et al. Agronomic values of greenwaste biochar as a soil amendment. Soil Research. 2008;**45**(8):629–634

[7] Dominic Woolf, James E. Amonette, F. Alayne Street-Perrott, Johannes Lehmann, Stephen Joseph. Sustainable biochar to mitigate global climate change. Nature Communications. 2010;**1**(3):118–124

[8] Abdullah H, Wu H. Biochar as a fuel: 1. Properties and grindability of biochars produced from the pyrolysis of mallee wood under slow-heating conditions. Energy & Fuels. 2009;**23**(8):4174–4181

[9] Uchimiya M, Lima I M, Thomas Klasson K, et al. Immobilization of heavy metal ions (CuII, CdII, NiII, and PbII) by broiler litter-derived biochars in water and soil. Journal of Agricultural and Food Chemistry. 2010;**58**(9):5538–5544

[10] Chen B, Chen Z, Lv S. A novel magnetic biochar efficiently sorbs organic pollutants and phosphate. Bioresource Technology. 2011;**102**(2):716–723

[11] Fang C, Zhang T, Li P, et al. Application of magnesium modified corn biochar for phosphorus removal and recovery from swine wastewater. International Journal of Environmental Research and Public Health. 2014;**11**(9):9217–9237

[12] Gale F, Marti D, Hu D. China's volatile pork industry. Situation & Outlook Report Rice, 2012;1–27

[13] The first National Census of Pollution Source Codification Committee. The first national source of pollution Census Information Collection IV: Pollution Source Census Technical Report. Beijing: China Environmental Science Press. (In Chinese)

[14] Xu X, Schierz A, Xu N, et al. Comparison of the characteristics and mechanisms of Hg (II) sorption by biochars and activated carbon. Journal of Colloid and Interface Science. 2016;**463**:55–60

[15] Han Y, Cao X, Ouyang X, et al. Adsorption kinetics of magnetic biochar derived from peanut hull on removal of Cr (VI) from aqueous solution: Effects of production conditions and particle size. Chemosphere. 2016;**145**:336–341

[16] Yao Y, Gao B, Inyang M, et al. Removal of phosphate from aqueous solution by biochar derived from anaerobically digested sugar beet tailings. Journal of Hazardous Materials. 2011;**190**(1):501–507

[17] Onay O, Kockar OM. Slow, fast and flash pyrolysis of rapeseed. Renewable Energy. 2003;**28**(15):2417–2433

[18] Kambo HS, Dutta A. A comparative review of biochar and hydrochar in terms of production, physico-chemical properties and applications. Renewable and Sustainable Energy Reviews. 2015;**45**:359–378

[19] Tan X, Liu Y, Zeng G, et al. Application of biochar for the removal of pollutants from aqueous solutions. Chemosphere. 2015;**125**:70–85

[20] Berge ND, Li L, Flora JRV, et al. Assessing the environmental impact of energy production from hydrochar generated via hydrothermal carbonization of food wastes. Waste Management. 2015;**43**:203–217

[21] Regmi P, Moscoso JLG, Kumar S, et al. Removal of copper and cadmium from aqueous solution using switchgrass biochar produced via hydrothermal carbonization process. Journal of Environmental Management. 2012;**109**(17):61–69

[22] Zhang ZB, Cao XH, Liang P, et al. Adsorption of uranium from aqueous solution using biochar produced by hydrothermal carbonization. Journal of Radioanalytical & Nuclear Chemistry. 2013;**295**(2):1201–1208

[23] Kumar S, Loganathan VA, Gupta RB, et al. An Assessment of U(VI) removal from groundwater using biochar produced from hydrothermal carbonization. Journal of Environmental Management. 2011;**92**(10):2504–2512

[24] Ahmad M, Rajapaksha AU, Lim JE, et al. Biochar as a sorbent for contaminant management in soil and water: A review. Chemosphere. 2014;**99**:19–33

[25] Grima-Olmedo C, Ramírez-Gómez Á, Gómez-Limón D, et al. Activated carbon from flash pyrolysis of eucalyptus residue. Heliyon. 2016;**2**(9):e00155

[26] Oliveira I, Blöhse D, Ramke HG. Hydrothermal carbonization of agricultural residues. Bioresource Technology. 2013;**142**(4):138–146

[27] Ci Fang, Tao Zhang, Rongfeng Jiang. Research advance of biochar adsorption for phosphorus recovery from wastewater. China Science Paper. 2015;**10**(3):309–315. (in Chinese)

[28] Sun H, Hockaday WC, Masiello CA, et al. Multiple controls on the chemical and physical structure of biochars. Industrial & Engineering Chemistry Research. 2012;**51**(9):3587–3597

[29] Fengfeng Ma, Baowei Zhao, Bin Nian. Adsorption of ammonia nitrogen in water by corn stalk biochar. Journal of Lanzhou Jiaotong University. 2015;**1**:024.(in Chinese)

[30] Beesley L, Moreno-Jiménez E, Gomez-Eyles JL, et al. A review of biochars' potential role in the remediation, revegetation and restoration of contaminated soils. Environmental Pollution. 2011;**159**(12):3269–3282

[31] Pagga U, Taeger K. Development of a method for adsorption of dyestuffs on activated sludge. Water Research. 1994;**28**(5):1051–1057

[32] Yao H, Lu J, Wu J, et al. Adsorption of fluoroquinolone antibiotics by wastewater sludge biochar: Role of the sludge source. Water, Air, & Soil Pollution. 2013;**224**(1):1–9

[33] Zheng H, Wang Z, Zhao J, et al. Sorption of antibiotic sulfamethoxazole varies with biochars produced at different temperatures. Environmental Pollution. 2013;**181**:60–67

[34] Xu R, Xiao S, Yuan J, et al. Adsorption of methyl violet from aqueous solutions by the biochars derived from crop residues. Bioresource Technology. 2011;**102**(22):10293–10298

[35] Sun L, Wan S, Luo W. Biochars prepared from anaerobic digestion residue, palm bark, and eucalyptus for adsorption of cationic methylene blue dye: Characterization, equilibrium, and kinetic studies. Bioresource Technology. 2013;**140**:406–413

[36] Qiu Y, Zheng Z, Zhou Z, et al. Effectiveness and mechanisms of dye adsorption on a straw-based biochar. Bioresource Technology. 2009;**100**(21):5348–5351

[37] Yang Y, Lin X, Wei B, et al. Evaluation of adsorption potential of bamboo biochar for metal-complex dye: equilibrium, kinetics and artificial neural network modeling. International Journal of Environmental Science and Technology. 2014;**11**(4):1093–1100

[38] Yinhai Lang, Wei Liu, Hui Wang. Study on the adsorption properties of biochar for pentachlorophenol in water. Environmental Science in China. 2014(8):2017–2023 (in Chinese)

[39] Zheng W, Guo M, Chow T, et al. Sorption properties of greenwaste biochar for two triazine pesticides. Journal of Hazardous Materials. 2010;**181**(1):121–126

[40] Nguyen MV, Lee BK. Removal of dimethyl sulfide from aqueous solution using cost-effective modified chicken manure biochar produced from slow pyrolysis. Sustainability. 2015;**7**(11):15057–15072

[41] Zhou F, Wang H, Zhang W, et al. Pb (II), Cr (VI) and atrazine sorption behavior on sludge-derived biochar: role of humic acids. Environmental Science and Pollution Research. 2015;**22**(20):16031–16039

[42] Wang Y, Lu J, Wu J, et al. Adsorptive Removal of Fluoroquinolone Antibiotics Using Bamboo Biochar. Sustainability. 2015;**7**(9):12947–12957

[43] Devi P, Saroha AK. Simultaneous adsorption and dechlorination of pentachlorophenol from effluent by Ni–ZVI magnetic biochar composites synthesized from paper mill sludge. Chemical Engineering Journal. 2015;**271**:195–203

[44] Yuehua Jin. Water pollution of heavy metals and its removal method. Environment and Life. 2014;**22**:136 (in Chinese)

[45] Inyang M, Gao B, Yao Y, et al. Removal of heavy metals from aqueous solution by biochars derived from anaerobically digested biomass. Bioresource Technology. 2012;**110**: 50–56

[46] Jiyi Zhang Liping Liang Lijun PuAdsorption of Cr (VI) on Wheat Straw Heat - treated Biochar. Journal of Lanzhou University of Technology. 2011;**37**(2):64–68 (in Chinese)

[47] Johansson CL, Paul NA, de Nys R, et al. Simultaneous biosorption of selenium, arsenic and molybdenum with modified algal-based biochars. Journal of Environmental Management. 2016;**165**:117–123

[48] Park JH, Ok YS, Kim SH, et al. Characteristics of biochars derived from fruit tree pruning wastes and their effects on lead adsorption. Journal of the Korean Society for Applied Biological Chemistry. 2015;**58**(5):751–760

[49] Zhang W, Zheng J, Zheng P, et al. Sludge-Derived Biochar for Arsenic(III) Immobilization: Effects of Solution Chemistry on Sorption Behavior. Journal of Environmental Quality, 2015;**44**(4):1119

[50] Zhang F, Wang X, Yin D, et al. Efficiency and mechanisms of Cd removal from aqueous solution by biochar derived from water hyacinth (*Eichornia crassipes*). Journal of Environmental Management. 2015;**153**:68–73

[51] Tytlak A, Oleszczuk P, Dobrowolski R. Sorption and desorption of Cr (VI) ions from water by biochars in different environmental conditions. Environmental Science and Pollution Research. 2015;**22**(8):5985–5994

[52] Yang Zhang, Zifu Li, Lin Zhang, et al. Study on adsorption of ammonia nitrogen by modified cob biochar. Journal of Chemical Industry and Engineering (China). 2014;**65**(3): 960–966 (in Chinese)

[53] Fengfeng Ma, Baowei Zhao, Jingru Diao, et al. Adsorption of ammonia nitrogen in water by biochar from cattle manure. Environmental Science. 2015;**5**:1678–1685 (in Chinese)

[54] Weifeng Cheng, Hui Li, Yanqin Yang, et al., Study on preparation of biochar and its adsorption of nitrogen and phosphorus by pyrolysis of anaerobic fermentation residue from municipal sludge. Acta Chimica Sinica. 2016;**4**:1541–1548 (in Chinese)

[55] Fang C, Zhang T, Li P, et al. Phosphorus recovery from biogas fermentation liquid by Ca–Mg loaded biochar. Journal of Environmental Sciences. 2015;**29**:106–114

[56] Verheijen F G A, Jeffery S, Bastos A C, et al. Biochar application to soils: A critical scientific review of effects on soil properties, processes and functions. Biochar Application to Soils - A Critical Scientific Review of Effects on Soil Properties, Processes and Functions. 2010.

[57] Zhang W, Mao S, Chen H, et al. Pb (II) and Cr (VI) sorption by biochars pyrolyzed from the municipal wastewater sludge under different heating conditions. Bioresource Technology. 2013;**147**:545–552